Lecture Notes in Artificial Intelligence 7589

Subseries of Lecture Notes in Computer Science

LNAI Series Editors

Randy Goebel
  University of Alberta, Edmonton, Canada
Yuzuru Tanaka
  Hokkaido University, Sapporo, Japan
Wolfgang Wahlster
  DFKI and Saarland University, Saarbrücken, Germany

LNAI Founding Series Editor

Joerg Siekmann
  DFKI and Saarland University, Saarbrücken, Germany

Leliane N. Barros
Marcelo Finger
Aurora T. R. Pozo
Gustavo A. Gimenénez-Lugo
Marcos Castilho (Eds.)

Advances in Artificial Intelligence - SBIA 2012

21th Brazilian Symposium on Artificial Intelligence
Curitiba, Brazil, October 20-25, 2012
Proceedings

 Springer

Series Editors

Randy Goebel, University of Alberta, Edmonton, Canada
Jörg Siekmann, University of Saarland, Saarbrücken, Germany
Wolfgang Wahlster, DFKI and University of Saarland, Saarbrücken, Germany

Volume Editors

Leliane N. Barros
Marcelo Finger
University of São Paulo
IME, Department of Computer Science, São Paulo, Brazil
E-mail: {leliane, mfinger}@ime.usp.br

Aurora T. R. Pozo
Marcos Castilho
Federal University of Paraná
Department of Informatics, Curitiba, Brazil
E-mail: aurora@inf.ufpr.br castilho@ufpr.br

Gustavo A. Gimenénez-Lugo
Federal University of Technology, Paraná, Curitiba, Brazil
E-mail: gustavo@dainf.ct.utfpr.edu.br

ISSN 0302-9743   e-ISSN 1611-3349
ISBN 978-3-642-34458-9   e-ISBN 978-3-642-34459-6
DOI 10.1007/978-3-642-34459-6
Springer Heidelberg Dordrecht London New York

Library of Congress Control Number: 2012949764

CR Subject Classification (1998): I.2, H.3, H.4, I.4, I.5, H.5

LNCS Sublibrary: SL 7 – Artificial Intelligence

Typesetting: Camera-ready by author, data conversion by Scientific Publishing Services, Chennai, India

Printed on acid-free paper

Springer is part of Springer Science+Business Media (www.springer.com)

Preface

After 16 years, the biannual Brazilian Symposium on Artificial Intelligence (SBIA) returned to Curitiba, Paraná, which also hosted its 13th edition in 1996.

Already in its 21st edition, SBIA has been consolidated as the central event on artificial intelligence (AI) in Brazil and an important international forum for discussion of new ideas and current projects for researchers, students, and practitioners in all areas of AI. This edition was held jointly with sister events SBRN, ENIA, and CTDIA under the umbrella of the Brazilian Conference of Intelligent Systems (BRACIS 2012).

These proceedings contain the 23 full papers selected for publication and oral presentation at SBIA 2012. There were 81 papers submitted from nine countries. The papers cover the following areas of AI with the corresponding number of accepted contributions:

- Knowledge representation: 5
- Machine learning: 7
- Machine learning and computer vision: 2
- Agent-based and multi-agent systems: 4
- Robotics and languages: 3
- Constraints: 2

We would like to thank all authors that submitted their work to SBIA 2012. We are also very grateful to the 134 members of our international Program Committee, who contributed with their knowledge, expertise, and time in doing an excellent job in reviewing the papers.

We are very thankful for all the support we received from the organization crew, in particular Aurora Pozo, BRACIS 2012 General Chair, and SBIA 2012 Local Chairs Gustavo Alberto Giménez Lugo and Marcos Castilho. A special thanks to the local universities that hosted and supported the organization of the event: UFPR, UTFPR, and PUCPR.

We thank the keynote speakers for accepting our invitation: Kenneth A. De Jong, Thomas G. Dietterich, Anthony Hunter, and Scott Sunner.

Finally, we thank SBIA 2012 sponsors CAPES, CNPq, Fapesp, and SBC for their support.

July 2012

Leliane Nunes Barros
Marcelo Finger

Organization

Organizing Committee

General Chair

Aurora T.R. Pozo — Universidade Federal do Paraná, Brazil

Local Chairs

Gustavo Alberto
  Giménez-Lugo — Universidade Tecnológica Federal do Paraná, Brazil

Marcos Castilho — Universidade Federal do Paraná, Brazil

Local Committee

Emerson Cabrera Paraiso — Pontifícia Universidade Católica do Paraná, Brazil

Júlio Cesar Nievola — Pontifícia Universidade Católica do Paraná, Brazil

Myriam Regattieri Delgado — Universidade Tecnológica Federal do Paraná, Brazil

Workshop Chair

Fabrício Enembreck — Pontifícia Universidade Católica do Paraná, Brazil

Tutorial Chairs

Alneu de Andrade Lopes — Universidade de São Paulo, Brazil
Roseli Aparecida Romero — Universidade de São Paulo, Brazil

Program Chairs

Leliane Nunes de Barros — Universidade de São Paulo, Brazil
Marcelo Finger — Universidade de São Paulo, Brazil

Steering Committee (CEIA)

Fabio Gagliardi Cozman — Universidade de São Paulo, Brazil
Augusto C.P. Loureiro
  da Costa — Universidade Federal da Bahia, Brazil
Antônio Carlos da Rocha
  Costa — Universidade Federal do Rio Grande, Brazil

Flavio Tonidandel	Centro Universitário da FEI, Brazil
Francisco Carvalho	Universidade Federal de Pernambuco, Brazil
Heloísa Camargo	Universidade Federal de São Carlos, Brazil
Jacques Wainer	Universidade Estadual de Campinas, Brazil

Program Committee

Adolfo Neto	Universidade Tecnológica Federal do Paraná, Brazil
Aline Villavicencio	Universidade Federal do Rio Grande do Sul, Brazil
Alipio Jorge	Universidade do Porto, Portugal
Alneu Lopes	Universidade de São Paulo, Brazil
Alvaro Moreira	Universidade Federal do Rio Grande do Sul, Brazil
Ana Bazzan	Universidade Federal do Rio Grande do Sul, Brazil
Ana Casali	Universidad Nacional de Rosario, Argentina
Ana Teresa Martins	Universidade Federal do Ceara, Brazil
Anarosa Brandão	Universidade de São Paulo, Brazil
Anderson de Araújo	Universidade de São Paulo, Brazil
André P. L. F de Carvalho	Universidade de São Paulo, Brazil
Andrea Omicini	Alma Mater Studiorum, Università di Bologna, Italy
Andreas Herzig	Institut de Recherche en Informatique de Toulouse, France
Anna Costa	Universidade de São Paulo, Brazil
Anne Canuto	Universidade Federal do Rio Grande do Norte, Brazil
Antônio Braga	Universidade Federal de Minas Gerais, Brazil
Antonio C. da Rocha Costa	Universidade Federal do Rio Grande, Brazil
Artur Garcez	City University of London, UK
Aurora Pozo	Universidade Federal do Paraná, Brazil
Benjamin Bedregal	Universidade Federal do Rio Grande do Norte, Brazil
Bianca Zadrozny	IBM Research, Brazil
Carlos Areces	Universidad Nacional de Córdoba, Argentina
Carlos Menezes	Universidade de São Paulo, Brazil
Carlos Ribeiro	Instituto Tecnológico da Aeronáutica, Brazil
Cassio de Campos	Dalle Molle Institute for Artificial Intelligence, Switzerland
Celso Kaestner	Universidade Tecnológica Federal do Paraná, Brazil
Christian Lemaitre	Universidad Autónoma Metropolitana, Mexico
Daniel Berrar	Tokyo Institute of Technology, Japan
Diana Adamatti	Universidade Federal do Rio Grande, Brazil

Donato Malerba	Università degli Studi di Bari, Italy
Edson Matsubara	Universidade Federal de Mato Grosso do Sul, Brazil
Edward Hermann Haeusler	Pontifícia Universidade Católica do Rio de Janeiro, Brazil
Eric Matson	Purdue University, USA
Eugenio Oliveira	Universidade do Porto, Portugal
Evandro Costa	Universidade Federal de Alagoas, Brazil
Fabiano Silva	Universidade Federal do paraná, Brazil
Fariba Sadri	Imperial College, UK
Fábio Cozman	Universidade de São Paulo, Brazil
Federico Barber	Universidad Politecnica de Valencia, Spain
Fernando Gomide	Universidade Estadual de Campinas, Brazil
Fernando Osorio	Universidade de São Paulo, Brazil
Filip Zelezny	Czech Technical University in Prague, Czech Republic
Flavio Tonidandel	Centro Universitario da FEI, Brazil
Flávio S. C. da Silva	Universidade de São Paulo, Brazil
Francesco Belardinelli	Imperial College of Science and Technology, UK
Francisco De Carvalho	Universidade Federal de Pernambuco, Brazil
Fred Freitas	Universidade Federal de Pernambuco, Brazil
Gabriel Lopes	Universidade Nova de Lisboa, Portugal
Geber Ramalho	Universidade Federal de Pernambuco, Brazil
Gerardo Schneider	University of Gothenburg, Sweden
Graça Gaspar	Universidade de Lisboa, Portugal
Gracaliz Dimuro	Universidade Federal do Rio Grande, Brazil
Guillermo Simari	Universidad Nacional del Sur in Bahia Blanca, Argentina
Gustavo A. Giménez-Lugo	Universidade Tecnológica Federal do Paraná, Brazil
Hector Geffner	Universitat Pompeu Fabra, Spain
Helder Coelho	Universidade de Lisboa, Portugal
Heloisa Camargo	Universidade Federal de são Carlos, Brazil
Ines Dutra	Universidade do Porto, Portugal
Ivandré Paraboni	Universidade de São Paulo, Brazil
Jaime Sichman	Universidade de São Paulo, Brazil
Jan Ramon	Katholieke Universiteit Leuven, Belgium
Joao Marcos	Universidade Federal do Rio Grande do Norte, Brazil
João Balsa	Universidade de Lisboa, Portugal
Jomi Hübner	Universidade Federal de Santa Catarina, Brazil
Juliana Bueno-Soler	Universidade Federal do ABC, Brazil
Julie Dugdale	MAGMA-Grenoble Informatics Laboratory, France
Karina Valdivia-Delgado	Universidade de São Paulo, Brazil

Kate Revoredo	Universidade Federal do Rio de Janeiro, Brazil
Krysia Broda	Imperial College, UK
Laurent Perrussel	Université Toulouse, France
Leliane Nunes de Barros	Universidade de São Paulo, Brazil
Leonardo Emmendorfer	Universidade Federal do Rio Grande, Brazil
Li Weigang	Universidade de Brasilia, Brazil
Luis Antunes	Universidade de Lisboa, Portugal
Luis Fariñas del Cerro	Université Paul Sabatier, France
Luis Lamb	Universidade Federal do Rio Grande do Sul, Brazil
Luis Otávio Alvares	Universidade Federal de Santa Catarina, Brazil
Luiz Satoru Ochi	Universidade Federal Fluminense, Brazil
Luiza Mourelle	Universidade do Estado do Rio de Janeiro, Brazil
Mara Abel	Universidade Federal do Rio Grande do Sul, Brazil
Marcelo Finger	Universidade de São Paulo, Brazil
Marcilio de Souto	Universidade Federal de Pernambuco, Brazil
Marcos Castilho	Universidade Federal do Paraná, Brazil
Maria Monard	Universidade de São Paulo, Brazil
Marilton Aguiar	Universidade Federal de Pelotas, Brazil
Mario Fernando Campos	Universidade Federal de Minas Gerais, Brazil
Marley M.B.R. Vellasco	Pontifícia Universidade Católica do Rio de Janeiro, Brazil
Márcio Ribeiro	Universidade Estadual de Campinas, Brazil
Mário Benevides	Universidade Federal do Rio de Janeiro, Brazil
Michael Thielscher	University of New South Wales, Australia
Nuno David	Instituto Universitário de Lisboa, Portugal
Olivier Boissier	ENS Mines Saint-Etienne, France
Patrícia Tedesco	Universidade Federal de Pernambuco, Brazil
Patricia Jaques	UNISINOS, Brazil
Patricia Oliveira	Universidade de São Paulo, Brazil
Paulo Ferreira Jr.	Universidade Federal de Pelotas, Brazil
Paulo Quaresma	Universidade de Évora, Portugal
Paulo Santos	Centro Universitário da FEI, Brazil
Paulo Trigo	Instituto Superior de Engenharia de Lisboa, Portugal
Petrucio Viana	Universidade Federal Fluminense, Brazil
Ramon de Mantaras	IIIA-CSIC, Spain
Reinaldo Bianchi	Centro Universitário da FEI, Brazil
Rejane Frozza	Universidade de Santa Cruz do Sul, Brazil
Renata de Freitas	Universidade Federal Fluminense, Brazil
Renata Vieira	Pontifícia Universidade Católica do Rio Grande do Sul, Brazil
Renata Wassermann	Universidade de São Paulo, Brazil

Ricardo Rabelo	Federal University of Santa Catarina, Brazil
Ricardo Silveira	Universidade Federal de Santa Catarina, Brazil
Rodrigo Braz	SRI International, USA
Ronaldo Prati	Universidade Federal do ABC, Brazil
Rosa Viccari	Universidade Federal do Rio Grande do Sul, Brazil
Rui Camacho	University of Porto, Portugal
Sandra Aluisio	Universidade de São Paulo, Brazil
Sandra Sandri	Instituto Nacional de Pesquisas Espaciais, Brazil
Sheila Veloso	Universidade do Estado do Rio de Janeiro, Brazil
Silvia Botelho	Universidade Federal do Rio Grande, Brazil
Silvia Schiaffino	Universidad Nacional de Buenos Aires, Argentina
Silvio Pereira	Faculdade Tecnológica de São Paulo, Brazil
Solange Rezende	Universidade de São Paulo, Brazil
Stefano Ferilli	Università di Bari, Italy
Teresa Ludermir	Universidade Federal de Pernambuco, Brazil
Thiago Pardo	Universidade de São Paulo, Brazil
Tiago de Lima	University of Lille Nord, France
Toby Walsh	NICTA and University of New South Wales, Australia
Torsten Schaub	University of Potsdam, Germany
Valerie Camps	Paul Sabatier University, France
Vania Bogorny	Universidade Federal de Santa Catarina, Brazil
Vítor Santos Costa	Universidade do Porto, Brazil
Viviane Torres da Silva	Federal University Fluminense, Brazil
Wagner Meira Jr.	Federal University of Minas Gerais, Brazil
Walter Carnielli	Universidade Estadual de Campinas, Brazil
Wamberto Vasconcelos	University of Aberdeen, UK
Yves Demazeau	Domaine Universitaire de Saint-Martin, France
Zhao Liang	University of São Paulo, Brazil

Supporting Scientific Society

SBC - Sociedade Brasileira de Computação

Sponsoring Institutions

CNPq - Conselho Nacional de Desenvolvimento Científico e Tecnológico
UFPR - Universidade Federal do Paraná
UTFPR - Universidade Tecnológica Federal do Paraná
PUCPR - Pontifícia Universidade Católica do Paraná

Table of Contents

Chapter 1: Knowledge Representation

Chapter 2: Machine Learning

Chapter 3: Machine Learning and Computer Vision

Chapter 4: Multi Agent Systems

Chapter 5: Robotic and Language

Chapter 6: Constraints

On the Development of a Formal Methodology for Knowledge Representation in Defeasible Logic Programming

Alejandro G. Stankevicius[1] and Marcela Capobianco[1,2]

[1] Artificial Intelligence Research and Development Laboratory
Department of Computer Science and Engineering
Universidad Nacional del Sur – Av. Alem 1253, (8000) Bahía Blanca, Argentina
[2] CONICET - Consejo Nacional de Investigaciones Científicas y Técnicas
{ags,mc}@cs.uns.edu.ar

Abstract. *Defeasible Logic Programming* (DeLP) is a formalism able to represent incomplete and potentially contradictory information that combines logic programming with defeasible argumentation. In the past few years, this formalism has been applied to real world scenarios with encouraging results. Not withstanding, the outcome one may obtain in this or any other argumentative system is directly related to the decisions (or lack thereof) made during the phase of knowledge representation. In addition, this is exacerbated by the usual lack of a formal methodology able to assist the knowledge engineer during this critical phase.

In this article, we propose a formal methodology for knowledge representation in DeLP, that defines a set of guidelines to be used during this phase. Our methodology results in an key tool to improve DeLP's applicability to concrete domains.

Keywords: knowledge representation, non-monotonic reasoning, defeasible reasoning, argumentative systems.

1 Introduction

Humans have always been intrigued by their ability to *reason*, even before the invention of computers. For decades, we have attempted to emulate the process of reasoning, but it proved to be quite hard to capture in a practical sense. Scientist have tried almost everything, from physiological explanations to sociological accounts. Out of those approaches, the one with possibly the longest tradition, going back at least to the golden age in classical Greece, conceives this process as a mere manipulation of *symbols* [10]. This great insight produced nowadays the most useful and versatile creations at our disposal (computers included). However, symbolic reasoning cannot be applied directly over the problem at hand: we require that the knowledge about a particular domain be also described symbolically, where this description is in turn the outcome of the process called *Knowledge Representation.*

L.N. Barros et al. (Eds.): SBIA 2012, LNAI 7589, pp. 1–11, 2012.

From a knowledge engineer's standpoint, knowledge representation is the process upon which we classify all the information present in a given domain into two categories: the core characteristics of the scenario under study, and the rest of the knowledge, that is, those other aspects deemed not relevant considering the problem at hand. Although this may look simple or straight, this is by no means trivial. Sometimes, it is more an art than a science, provided that what can be derived applying this kind of reasoning is usually intrinsically intertwined with the way knowledge is being represented. When dealing with this, what we actually require is a set of guidelines assuring us that no core characteristics may mistakenly be ignored. In other words, what we need is a *methodology*.

Defeasible Logic Programming (DeLP) is a formalism able to represent incomplete and potentially contradictory information that combines logic programming with defeasible argumentation. Ideas borrowed from defeasible argumentation such as representing defeasible reasons as arguments, or performing a full dialectical analysis before answering queries are incorporated into a knowledge representation language featuring PROLOG-like rules. This formalism has recently been applied to real world problems with encouraging results [6,5]. Not withstanding, the outcome one may obtain in this or any other argumentative system is directly related to the decisions (or lack thereof) made during the phase of knowledge representation. In addition, this is exacerbated by the usual lack of a formal methodology able to assist the knowledge engineer during this critical phase. This problem is common in argumentation systems, given that no formal knowledge representation methodologies have been proposed for such formalisms that could be used to standarize the modelling process of a given domain.

In this article, we propose a formal methodology with the intent of standarizing the process of knowledge representation in DeLP which encompasses a set of guidelines to be used during this key phase. We hold that having a methodology is crucial in order to improve DeLP's applicability to concrete domains. The work reported here refines and extends a previous work where a preliminary version of this methodology was initially published [13].

In what follows, Section 2 discusses related work on other methodologies for knowledge representation. Section 3 briefly introduces the inner working of DeLP. Then, Section 4 outlines the proposed methodology. Section 5 reviews some knowledge representation properties in the context of DeLP. Finally, Section 6 presents the conclusions of the article.

2   Knowledge Representation Methodologies

The logic programming community has made significant contributions on methodologies for knowledge representation in non-monotonic formalisms. For instance, Baral and Gelfond's survey [2] discussed alternative ways of representing knowledge using general logic programs, extended logic programs and disjunctive logics programs. They analyzed, among other topics, how to formalize normative statements, the closed world and domain closure assumptions and other types of defaults and autoepistemic statements.

Gelfond and Leone [8] analyzed knowledge representation in the context of a family of declarative logic program languages based on answer set/stable models semantics called A-Prolog. This paper relates the development of a working example illustrating some interesting features of A-Prolog, such as recursive rules, the use of defaults with exceptions or the ability to distinguish between falsity and the absence of informations. It also includes the analysis of several interesting concepts such as elaboration tolerant databases, the representation of different types of negation and its implications, and encoding defaults with exceptions.

In his book on Answer set logic programming (AnsProlog), Chitta Baral [1] addresses the problem of obtaining a complete knowledge representation methodology for that formalism. He presents several examples corresponding to different declarative problem solving modules or knowledge representation and reasoning aspects. Baral also shows the equivalence of AnsProlog with respect to other widely known non-monotonic formalisms, analyzing the properties of restrict monotonicity, language independence, language tolerance and compact representation in the context of AnsProlog.

Regarding the argumentation community, it should be emphasized that no argumentation formalism has currently been provided with a methodology thorough enough. We intend to follow the same path Baral, Gelfond and Leone traveled in the logic programming community in the context of argumentation systems.

3 Defeasible Logic Programming

In this section we briefly review the key definitions of the DeLP system as presented in [7]. These concepts are needed later in Section 4 in order to develop our methodology.

The language of DeLP is based on the language of logic programming. Standard logic programming concepts (such as signature, variables, functions, etc) are defined in the usual way. Literals are atoms that may be preceded by the symbol "$\sim$" denoting *strict* negation, as in extended logic programming. Facts are simply literals. Strict rules are ordered pairs $L_0 \leftarrow L_1, \ldots, L_n$ whose first component, L_0, is a literal, and whose second component, $L_1, \ldots, L_n$, is a finite non-empty set of literals. Defeasible rules are ordered pairs $L_0 \prec L_1, \ldots, L_n$ whose first component, L_0, is a literal, and whose second component, $L_1, \ldots, L_n$, is a finite non-empty set of literals. Defeasible rules are used to represent defeasible knowledge (*i.e.*, tentative information that can be used as long as nothing is posed against it), whereas strict rules are used to represent incontrovertible information.

In this formalism, the state of the world is modelled as a *Defeasible Logic Program (de.l.p)*, essentially a possibly infinite set of facts, strict rules and defeasible rules. In a given *de.l.p* $\mathcal{P}$, the subset of facts and strict rules is referred to as Π, and the subset of defeasible rules as Δ, thus a *de.l.p* $\mathcal{P}$ can also be noted as (Π, Δ). Since the set Π represent non-defeasible information it must be non-contradictory.

Given a *de.l.p* program $\mathcal{P}$, a query posed to $\mathcal{P}$ is a ground literal Q which must be supported by an *argument*. Arguments are built on the basis of a *defeasible derivation* computed by backward chaining applying the usual SLD inference procedure used in logic programming.

Definition 1. Argument structure*[7]Let h be a literal, and $\mathcal{P} = (\Pi, \Delta)$ be a de.l.p. We say that $\langle \mathcal{A}, h \rangle$ is an* argument structure *for h, if, and only if, $\mathcal{A}$ is a set of defeasible rules from $\mathcal{P}$ (i.e., $\mathcal{A} \subseteq \Delta$), such that: (1) there exists a defeasible derivation for h from $\Pi \cup \mathcal{A}$, (2) the set $\Pi \cup \mathcal{A}$ is non-contradictory, and (3) $\mathcal{A}$ is minimal with respect to set inclusion (i.e., no $\mathcal{A}' \subset \mathcal{A}$ satisfies the previous conditions).*

Arguments can attack each other, an argument $\langle \mathcal{A}_1, h_1 \rangle$ *attacks* $\langle \mathcal{A}_2, h_2 \rangle$ at h if, and only if, there exists a sub-argument structure $\langle A, h \rangle$ from $\langle \mathcal{A}_2, h_2 \rangle$ such that $\Pi \cup \{h, h_1\}$ is contradictory. Defeat among arguments is defined combining the attack relation and a preference criterion "$\preceq$". An argument $\langle \mathcal{A}_1, h_1 \rangle$ *defeats* $\langle \mathcal{A}_2, h_2 \rangle$ if $\langle \mathcal{A}_1, h_1 \rangle$ attacks $\langle \mathcal{A}_2, h_2 \rangle$ at a literal h and $\langle \mathcal{A}_1, h_1 \rangle \preceq \langle A, h \rangle$ (proper defeater) or $\langle \mathcal{A}_1, h_1 \rangle$ is unrelated to $\langle A, h \rangle$ (*blocking defeater*).

Defeaters are arguments and may in turn be defeated. Thus, a complete dialectical analysis is required to determine which arguments are ultimately accepted. Such analysis results in a tree structure called *dialectical tree*, in which arguments are nodes labeled as undefeated (U-nodes) or defeated (D-nodes) according to a marking procedure.

Definition 2. Dialectical tree *[7]The* dialectical tree *for an argument $\langle \mathcal{A}, h \rangle$, denoted $\mathcal{T}_{\langle \mathcal{A}, h \rangle}$, is recursively defined as follows: (1) A single node labeled with an argument $\langle \mathcal{A}, h \rangle$ with no defeaters (proper or blocking) is by itself the dialectical tree for $\langle \mathcal{A}, h \rangle$; (2) Let $\langle \mathcal{A}_1, h_1 \rangle, \langle \mathcal{A}_2, h_2 \rangle, \ldots, \langle \mathcal{A}_n, h_n \rangle$ be all the defeaters (proper or blocking) for $\langle \mathcal{A}, h \rangle$. The dialectical tree for $\langle \mathcal{A}, h \rangle$, $\mathcal{T}_{\langle \mathcal{A}, h \rangle}$, is obtained by labeling the root node with $\langle \mathcal{A}, h \rangle$, and making this node the parent of the root nodes for the dialectical trees of $\langle \mathcal{A}_1, h_1 \rangle, \langle \mathcal{A}_2, h_2 \rangle, \ldots, \langle \mathcal{A}_n, h_n \rangle$.*

For the marking procedure we start labeling the leaves as U-nodes. Then, for any inner node $\langle \mathcal{A}_2, Q_2 \rangle$, it will be marked as U-node if and only if every child of $\langle \mathcal{A}_2, Q_2 \rangle$ is marked as a D-node. If $\langle \mathcal{A}_2, Q_2 \rangle$ has at least one child marked as U-node then it is marked as a D-node. This marking allows us to characterize the set of literals sanctioned by a given *de.l.p*, called *warranted literals*. A literal h is *warranted* if and only if there exists an argument structure $\langle \mathcal{A}, h \rangle$ for h, such that the root of its marked dialectical tree $\mathcal{T}_{\langle \mathcal{A}, h \rangle}^{\star}$ is a U-node.

Example 1. The *de.l.p* in Figure 1 represents information about felines and the dialectical tree at the right explains the answer to the query climbs_tree(simba) with respect to this program.

4   Development of a Formal Methodology

This section proposes a set of guidelines describing how different situations involving non-monotonic knowledge, incomplete information, and conflicting information can be modelled within DeLP.

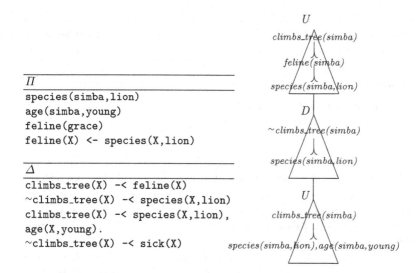

Fig. 1. A *de.l.p* program about felines

Relational Databases. To begin with, the most essential form of symbolic knowledge one may want to represent is a set atomic formulas, where in turn each formula states that a certain relation holds over a given tuple of objects. This kind of knowledge can be represented in DeLP as a set of facts, much in the same way this knowledge is represented in PROLOG. Bear in mind that the nature of these relations, or the origin of the objects they relate cannot be inspected: it should be determined by the knowledge engineer for each concrete scenario.

Remark 1. Let r be a finite relation over a certain domain. This relation can be modelled in DeLP by adding a new fact $r(\overline{X})$ for every $\overline{X} \in r$.

The reader might notice only finite relations can be modelled. Then again, the same holds for relational databases and PROLOG programs.

Example 2. $\mathcal{P} = (\Pi, \emptyset)$, where $\Pi = \{\texttt{weekend(saturday)}; \texttt{weekend(sunday)}\}$, models the relation 'weekend day' defined over the days of the week.

Class Inclusion. Another important piece of knowledge usually worth modelling is the inclusion of concepts or categories. From Aristotle's syllogisms to modern day inheritance hierarchies, being able to represent class inclusion has always been deemed crucial. The classical approach consists in modelling the inclusion of a concept A into another concept B with a material implication $A(\overline{X}) \rightarrow B(\overline{X})$, denoting that all the instances of the former are also instances of the latter. We can follow a similar approach in DeLP, as long as we take into consideration that DeLP does not have material implication but inference rules (meta-relations between set of formulas to be precise). That is:

Remark 2. Let A and B be two concepts, already modelled as relations among objects, such that A is included in B. This knowledge can be expressed in a *de.l.p* by adding the strict rules $b(\overline{X}) \leftarrow a(\overline{X})$ and $\sim a(\overline{X}) \leftarrow \sim b(\overline{X})$.

Much to our surprise, this result made us aware that all this time we have been modelling class inclusion the wrong way. Considering that DeLP is a refinement of Simari-Loui's system [12], where strict rules were material implications, we incorrectly kept modelling this notion with the first rule alone, when in fact two rules were actually required.

Example 3. $\mathcal{P} = (\Pi, \emptyset)$, where $\Pi = \{$mammal(X) <- human(X);
~human(X) <- ~mammal(X)$\}$, models the fact that humans are mammals.

Negative Information. So far, only positive information was considered. Even though positive information is our preferred source of knowledge, sometimes negative information constitutes a valid source of additional knowledge. Both relational databases and PROLOG adopt some sort of Closed-World Assumption (CWA) in order to represent this kind of information, but, by doing so, a subtle aspect is lost in the process: CWA does not distinguish between not knowing whether a certain relation holds over some objects, and knowing that this relation in fact does not hold over those objects. This issue has been addressed in PROLOG giving birth to *extended logic programming* [9], a formalism that extends PROLOG allowing the explicit representation of negative information. Following a similar approach, DeLP also allows the representation of this kind of knowledge.

Remark 3. Let r be a relation and $\overline{X}$ a set of objects such that we have witnessed that $\overline{X} \notin r$. This knowledge can be expressed in a *de.l.p* by adding a new fact $\sim r(\overline{X})$.

Lack of Information. DeLP can represent both positive and negative information, yet we cannot explicitly encode lack of information. However, there is an extension of DeLP which incorporates default negation also reported in [7]. This kind of negation is only allowed preceding literals in the body of defeasible rules. The intended meaning is that the default negation of some literal q, noted **not** q, should be interpreted as *there is no warrant for q*. Default negation allow us to represent this lack of information as follows:

Remark 4. Let q be a relation about which we want to state that it is not known whether q holds over some tuple $\overline{X}$ This knowledge can be expressed in a *de.l.p* by adding a new defeasible rule $unknown(q(\overline{X})) \prec \mathbf{not}\ q(\overline{X}), \mathbf{not}\ \sim q(\overline{X})$.

Example 4. Consider a *de.l.p* $\mathcal{P} = (\emptyset, \Delta)$ describing the eligibility condition for a certain fellowship, with $\Delta = \{$eligible(X) -< fairGPA(X), special(X);
eligible(X) -< highGPA(X); ~eligible(X) -< ~special(X),~highGPA(X);
interview(X) -< not eligible(X), not ~eligible(X)$\}$. Observe the last rule models that when we do not know whether a particular applicant is eligible or not, an additional interview with him or her should be conducted.

Database Constraints. The conflict arising from complementary literals, called *explicit conflict* (for instance, "being alive" and "not being alive"), is a situation that DeLP's inference engine captures by itself. However, sometimes the conflict existing between a set of situations is more subtle (*i.e.*, less syntactic), and does not involve complementary concepts (for instance, "being alive" and "being dead"). In a sense, this conflict represents a set of situations that cannot happen all at the same time, that is, it is a kind of *constraint*.

Remark 5. Let $S_1, S_2, \ldots, S_n$ be all the situations characterizing an explicit conflict. This constraint can be expressed in a *de.l.p* by adding n new strict rules of the form $\sim S_i \leftarrow S_1, \ldots, S_{i-1}, S_{i+1}, \ldots, S_n$, for every S_i, $1 \leq i \leq n$.

Example 5. $\mathcal{P} = (\Pi, \emptyset)$, with $\Pi = \{\sim\texttt{dead(X)} \texttt{ <- alive(X)}; \sim\texttt{alive(X) <- dead(X)}\}$, captures the constraint that one cannot be alive and dead at the same time.

Defaults. When it comes to make the case for non-monotonic reasoning, most formal theories resort to argue that there is a feature of common-sense reasoning which cannot be captured using monotonic theories: *defaults*. A default establishes a connection between concepts weaker than class inclusion, yet relevant enough as to require to be made explicit. Put the other way around, defaults can be used to model *relations with exceptions*. The classical example in Artificial Intelligence is "birds usually fly". Observe that this relation cannot be expressed as class inclusion (*e.g.*, "birds fly"), because it becomes inconsistent with the fact that some birds do not fly (*e.g.*, penguins, ostriches, kiwis, etc.), nor it can be expressed as a rule with explicit exceptions (*e.g.*, "birds that are not penguins, kiwis, ... fly"), because all the exception to this relation must be known beforehand (an inconvenience known as the *qualification problem*). It is in this particular regard that DeLP really shines. Its representation of defaults is quite natural, and even more straightforward than the representation of defaults in Default Logic or other non-monotonic theories.

Remark 6. Let A and B be two situations such that when A occurs, B usually also occurs (that is to say, if A, then by default B). This knowledge can be expressed in a *de.l.p* simply by adding a new defeasible rule $B \prec A$.

Weak Constraint. Sometimes, explicit conflicts and constraints are not enough to capture all the conflicts worth modelling. For instance, there is a weaker form of conflict not covered so far, which we will call *weak constraint*. This kind of constraint reflects that two or more situations *usually* cannot occur at the same time, but *exceptionally*, they may. We model weak constraint like a database constraint, using defeasible rules instead of strict ones.

Remark 7. Let $S_1, S_2, \ldots, S_n$ be all the situations characterizing a weak conflict. This knowledge can be expressed in a *de.l.p* by adding n new defeasible rules of the form $\sim S_i \prec S_1, \ldots, S_{i-1}, S_{i+1}, \ldots, S_n$, for every S_i, $1 \leq i \leq n$.

Note that it is entirely possible to conceive a scenario where all these situations are met at the same time, considering the weak nature of this conflict, whereas strong conflicts make such a scenario outright impossible. It is the existence of this exceptional scenario what distinguishes weak from standard constraints.

Example 6. $\mathcal{P} = (\emptyset, \Delta)$, where $\Delta = \{\sim\texttt{sunny -< rainy}; \sim\texttt{rainy -< sunny}\}$, captures the weak constraint between a sunny day and the fact that it is raining.

Assumptions. Some theories incorporate non-monotonic features by allowing the reasoner to make tentative assumptions. This possibility alone gives rise to non-monotonic reasoning, since an assumption compatible under certain state of the world may become unfeasible with the addition of new information. For example, Bondarenko *et al.* [3] developed a formalisms built around abductive reasoning following this approach. Although the non-monotonic features in DeLP are the consequence of the dialectical analysis performed to determine if a conclusion is warranted, there is a particular kind of assumption, called *presumption*, which can also be modelled in this formalism. Bluntly put, a presumption is a *default fact*. This notion, first mentioned by Nute in his Defeasible Logic [11], allows the knowledge engineer to define a set of assumptions or suppositions which may be freely used as long as no other reasons are raised against them.

Remark 8. Let A be an assumption worth being modelled. This knowledge can be captured in a *de.l.p* by adding a new fact F, for a literal F not appearing in that *de.l.p*, and a new defeasible rule of the form $A \prec F$.

Example 7. Consider the following *de.l.p*:

Π	Δ
`bird(X) <- chicken(X)`	`flies(X) -< bird(X)`
`~chicken(X) <- ~bird(X)`	`~flies(X) -< chicken(X)`
`chicken(coco)`	`flies(X) -< chicken(X), scared(X)`
`assumption`	`scared(coco) -< assumption`

According to this *de.l.p*, *flies(coco)* is warranted (a conclusion reached under the assumption that *coco* was scared). Observe how the addition of the fact *assumption* and the corresponding defeasible rule captured the essence of the assumption that *coco* was scared.

5   DeLP as a Knowledge Representation Language

In this section we analyze how DeLP behaves with respect to the property of elaboration tolerance and show how DeLP relates with traditional logic programming.

Elaboration Tolerance. A good representation of defaults should allow the incorporation of information about exceptional elements with respect to these defaults . Elaboration tolerance meassures how straightforward is the incorporation of new information, that is, what classes of new information can be incorporated through local changes to the original representation. Consider $\mathcal{P}=(\Pi, \Delta)$, with $\Pi = \{$`feline(X) <- species(X,lion)`; `feline(simba)`$\}$ and the set of defeasible rules $\Delta = \{$`climbs_tree(X) -< feline(X)`$\}$. This program encodes the default "Felines usually climb trees". This representation is elaboration tolerant to new knowledge of the kind "Lions are exceptional felines that do not climb trees", there is no need to change the program to add this kind of knowledge, we only have to add a new rule to Δ: `~climbs_tree(X) -< species(X,lion)`. If we learnt next that Simba is a lion, a new fact `lion(simba)` should be added to Π and without changes in the other rules of the program we can now conclude that Simba cannot climb trees.

We could even add exceptions for exceptional defaults without making changes to the represented knowledge. Suppose we learn now that Simba is a cub and that cubs are exceptional to the default establishing that lions do not climb trees. We could add a new rule `climbs_tree(X) -< species(X,lion), young(X))` to Δ and a new fact `young(simba)` to Π. Then the system could conclude that Simba can climb trees without making any changes in the previous database. This simple example shows that DeLP can encode elaboration tolerant database is an seamless way.

Logic Programming. The paradigm of logic programming has been throughly explored as a tool for knowledge representation almost since its initial conception. Many attractive expert systems have been developed in PROLOG, the most well-known exponent of this paradigm. In these systems, the knowledge is represented through a standard PROLOG program. These programs, by extending a finite set of facts with general rules, allow the representation of infinite relations, something not possible using standard relational databases. These kind of knowledge can also be expressed in DeLP, since any standard PROLOG program can be reformulated as a *de.l.p* as follows:

Remark 9. Let P be a definite PROLOG program. The knowledge represented by the program P can be expressed as a *de.l.p* by virtue of the same set of facts and (strict) rules.

The previous remark rests upon the following proposition, relating the answers a given PROLOG program returns with DeLP's notion of warranted literals.

Proposition 1. *Let P be a definite* PROLOG *program. Then, q is a ground query entailed from P in* PROLOG *if, and only if, q is a warranted literal from (Π, Δ), where $\Pi = P$ and $\Delta = \emptyset$.*

Proof.   ($\Rightarrow$)
If q is entailed from P, there exist a SLD-derivation of q from P. Let $L_1, L_2, \ldots, L_n = q$ be the sequence of literals composing that SLD-derivation. Observe that this same sequence can also be used to construct a strict derivation of q from P, provided the same restrictions apply to both notions.[1] Therefore, $\langle \emptyset, q \rangle$ is a valid argument structure for q, if we take into account that q can be defeasibly derived from $P \cup \emptyset$, $P \cup \emptyset$ is non-contradictory since PROLOG program cannot express negative information, and that there are no subsets of $\emptyset$. This kind of argument structures (those based on strict derivations) are quite particular. For instance, no other argument structure can defeat them, no matter the argument-comparison criterion considered (this follows from Prop. 3.1 in [7]). That is to say, its corresponding dialectical tree $\mathcal{T}_{\langle \emptyset, q \rangle}$ has only one node labelled $\langle \emptyset, q \rangle$ and marked as U-node which in turn means that q is a warranted literal of this *de.l.p.*

                                                                                                          ($\Leftarrow$)
If q is a warranted literal from P, there must exist an argument structure $\langle \mathcal{A}, q \rangle$ for q. According to Def. 1, $\mathcal{A} \subset \Delta$, whereas $\Delta = \emptyset$. Therefore, $\mathcal{A} = \emptyset$, which means that there must exist a strict derivation of q from P. Let $L_1, L_2, \ldots, L_n = q$ be the

---

[1] The formal demonstration of this statement, by virtue of structural induction, is trivial.

sequence of literals composing that strict derivation. Once again, this same sequence can also be used to construct a SLD-derivation of q from P. Finally, since there exists a SLD-derivation of q from P, q must be entailed from P.

6   Conclusions

Defeasible Logic Programming constitutes an attractive framework for representing incomplete and potentially contradictory information, a cornerstone of many real-world problems. As such, the task of representing the knowledge required to reason upon these problems is not trivial. To that end, in this article we have developed a methodology for representing knowledge in DeLP, essentially a set of guidelines covering the major challenges knowledge engineers usually encounter during this phase. We have also shown the relation between DeLP and traditional logic programs and studied the property of elaboration tolerance in the context of DeLP.

As a future work, we plan to extend this methodology to cover other forms of monotonic and non-monotonic knowledge representation. For instance, we would like to further investigate the different ways of modelling priorities among defaults, formalizing causal effects of actions, and also take into account other variants of DeLP such as ODeLP [4].

Acknowledgements. This research was partially supported by CONICET (Argentina), and by the *Universidad Nacional del Sur*. The authors would also like to thank the anonymous reviewers for their valuable comments and suggestions that helped improve the quality of this work.

References

1. Baral, C.: Knowledge representation, reasoning and declarative problem solving. Cambridge University Press (2003)
2. Baral, C., Gelfond, M.: Logic programming and knowledge representation. Journal of Logic Programming 1(19-20), 73–148 (1994)
3. Bondarenko, A., Dung, P.M., Kowalski, R.A., Toni, F.: An abstract, argumentation-theoretic approach to default reasoning. Artificial Intelligence 93(1-2), 63–101 (1997)
4. Capobianco, M., Chesñevar, C.I., Simari, G.R.: Argumentation and the Dynamics of Warranted Beliefs in Changing Environments. Journal of Autonomous Agents and Multiagent Systems 11, 127–151 (2005)
5. Chesñevar, C., Maguitman, A., González, M.P.: Empowering recommendation technologies through argumentation. In: Argumentation in Artificial Intelligence. Springer (2010)
6. Chesñevar, C., Maguitman, A., Simari, G.: Argument-based critics and recommenders: A qualitative perspective on user support systems. Data & Knowledge Engineering 59(2), 293–319 (2006)
7. García, A.J., Simari, G.R.: Defeasible Logic Programming: An Argumentative Approach. Theory and Practice of Logic Programming 4(1), 95–138 (2004)

8. Gelfond, M., Leone, N.: Logic programming and knowledge representation - the a-prolog perspective. Artificial Intelligence 1-2(138), 3–38 (2002)
9. Gelfond, M., Lifschitz, V.: Logic Programs with Classical Negation. In: Warren, D.H.D., Szeredi, P. (eds.) Proceedings of the 7th International Conference on Logic Programming, pp. 579–597 (1990)
10. Newell, A., Simon, H.: Computer Science as Empirical Inquiry. In: ACM Turing Award Lectures, pp. 287–387. Addison-Wesley (1987)
11. Nute, D.: Defeasible Reasoning. In: Proceedings of the XX Annual Hawaii International Conference on System Sciences, pp. 470–477 (1987)
12. Simari, G.R., Loui, R.P.: A Mathematical Treatment of Defeasible Reasoning and its Implementation. Artificial Intelligence 53(1–2), 125–157 (1992)
13. Stankevicius, A.G., Simari, G.R.: A Methodology for Knowledge Representation in Defeasible Logic Programming. In: Proceedings del 12vo Congreso Argentino de Ciencias de la Computación (CACiC), pp. 1139–1150. Universidad Nacional de San Luis, San Luis (2006)

A Framework for Empirical Evaluation of Belief Change Operators

Renato U. Lundberg[1], Márcio M. Ribeiro[2], and Renata Wassermann[1]

[1] Instituto de Matemática e Estatística
Universidade de São Paulo
{lundberg,renata}@ime.usp.br
[2] Centro de Lógica e Epistemologia
Universidade Estadual de Campinas
marcio@cle.unicamp.br

Abstract. Belief revision has been extensively studied in the last thirty years. While there are many results in the literature comparing different operators from a theoretical point of view, there is no study of how the different operators perform in practice. In this paper, we propose a framework for empirical testing of belief change operators. The idea is that any operator can be quickly implemented making use of the available API and then tested for multiple scenarios. We illustrate the use of the framework with a case study comparing partial meet and kernel contraction operators.

1 Introduction

Belief Revision (or Belief Change) [1, 2] deals with the problem of incorporating new information to an existing knowledge base. Throughout the last three decades several different belief change operators were proposed. Usually these operators are described by means of *rationality postulates* as well as mathematical constructions. *Representation theorems* connect sets of postulates to constructions, allowing for theoretical comparisons between the different operators. Looking at the postulates and theorems, one can select constructions that seem intuitive for a given problem.

Much improvement has been made to compare operators, but there seems to be a gap when it comes to the computational resources required by them. There are many ways to execute a belief change, but how to choose the fastest given a specific problem instance?

When the main operators were proposed, in the 80s and 90s, it was not feasible to run extensive tests for lack of computational power. Although this problem still takes considerable time on hard instances, there are enough resources now to execute massive amounts of tests to analyse empirically the performance of each operator side by side.

In this work a computational framework for implementing and testing belief revision operators is proposed, aiming to provide an homogeneous ground to allow fair comparisons between them.

L.N. Barros et al. (Eds.): SBIA 2012, LNAI 7589, pp. 12–21, 2012.

There have been several implementations of belief revision operators in the literature, such as Immortal [3], BReLS [4] and SATEN [5]. However, each of them focusses on one particular family of constructions and logic. To our knowledge, there is no system intended to give general support as ours.

The paper proceeds as follows: in the next Section, we quickly present the necessary background on belief revision. In Section 3, we present the general structure of the proposed framework and in Section 4 we show how the different operators are implemented in the framework. Then we present in Section 5 an example of use of the framework for comparing two different operators. And finally, in 6 we summarize our contribution and point towards future work.

2   Background

Belief revision is a sub-field of knowledge representation that deals with the belief dynamics. *Operations* in belief revision theory model how an agent changes his beliefs.

There are many ways to represent the epistemic state of an agent. Representing it as a belief set (i.e. as a sets of sentences K closed under logical consequence $K = Cn(K)$) is the most studied way. In the seminal paper [6] the authors present three operations over belief sets. *Expansion* $(K + \alpha)$ consists simply in adding a new piece of information to the agent's epistemic state. *Revision* $(K * \alpha)$ also adds a new piece of information to the agent's epistemic state, but in a consistent way. *Contraction* $(K - \alpha)$ consists in removing beliefs from the agent's epistemic state. Both revision and contraction are defined via a set of rationality postulates (AGM postulates).

Many constructions for AGM contraction can be found in the literature [7–10]. Each of them is proved equivalent to the AGM postulates for contraction and can be used to define an AGM revision using the Levi identity $K * \alpha = (K - \neg\alpha) + \alpha$.

Another way for representing the epistemic state of an agent is via an arbitrary set of sentences B called *belief bases* [2]. Belief bases have certain advantages over belief sets such as being more expressive. For example belief bases $B = \{p, q\}$ and $B' = \{p \wedge q\}$ are distinct although their logical closure is the same. Furthermore, it clearly has advantages in computational complexity. For this latter reason this representation will be considered here.

Belief base operations also includes expansion, revision and contraction. Expansion in belief bases is simply defined as $B + \alpha = B \cup \{\alpha\}$. Given a contraction, a revision can be obtained via Levi identity $((B - \neg\alpha) + \alpha)$ or via reversed Levi identity $(B + \alpha - \neg\alpha)$ [11]. We will present in further subsections two ways of constructing a contraction: kernel contraction and partial meet contraction.

2.1   Kernel Contraction

Kernel contraction consists in finding the minimal subsets of B that imply the input α and removing from B at least one element of each such set. The set of all minimal subsets of B that imply α is called the *kernel* of B w.r.t. α $(B \perp\!\!\!\perp \alpha)$ and each element of it is called an α-*kernel*. Formally:

Definition 1 (Kernel). *A set X is an element of $B \perp\!\!\!\perp \alpha$ iff:*

1. $X \subseteq B$ *(X is a subset of B).*
2. $\alpha \in Cn(X)$ *(which implies α).*
3. *if $X' \subset X$ then $\alpha \notin Cn(X')$ (and is minimal)*

An *incision function* is a function σ that "chooses" at least one element of each α-kernel. Formally:

Definition 2 (incision function). *An* incision function *σ for belief base B is a function such that for every α:*

1. $\sigma(B \perp\!\!\!\perp \alpha) \subseteq \bigcup B \perp\!\!\!\perp \alpha$.
2. *if $\emptyset \neq X \in B \perp\!\!\!\perp \alpha$ then $X \cap \sigma(B \perp\!\!\!\perp \alpha) \neq \emptyset$.*

The *kernel contraction* $-_\sigma$ over a belief base B is defined as follows: $B -_\sigma \alpha = B \setminus \sigma(B \perp\!\!\!\perp \alpha)$.

The following postulates fully characterize the kernel contraction i.e. kernel contraction satisfies each of these postulates and any operation – that satisfies these postulates can be constructed as a kernel contraction [12].

(success) if $\alpha \notin Cn(\emptyset)$ then $\alpha \notin Cn(B - \alpha)$.
(inclusion) $B - \alpha \subseteq B$.
(uniformity) If for every $B' \subseteq B$ it holds that $\alpha \in Cn(B')$ iff $\beta \in Cn(B')$ then $B - \alpha = B - \beta$.
(core-retainment) If $\beta \in B \setminus B - \alpha$ then there is $B' \subseteq B$ such that $\alpha \notin Cn(B')$, but $\alpha \in Cn(B' \cup \{\beta\})$

Success guarantees that the input α is removed from the consequences of the belief base. *Inclusion* guarantees that no sentence is added to the base during the contraction. *Uniformity* guarantees that if two sentences α and β are consequences of the same subsets of B then the contraction $B - \alpha$ and $B - \beta$ should give the same result. Finally, *core-retainment* is a minimality criterion. It guarantees that only sentences that somehow helps to imply α should be removed from B.

2.2   Partial Meet Contraction

Partial meet contraction consists in finding the maximal subsets of B that do *not* imply the input α and then taking the intersection of some of these sets. The set of all maximal subsets of B that do not imply α is called the *remainder set* of B w.r.t. α $(B \perp \alpha)$. Formally:

Definition 1 (remainder set). *A set X is an element of $B \perp \alpha$ iff:*

1. $X \subseteq B$ *(X is a subset of B).*
2. $\alpha \notin Cn(X)$ *(that doesn't imply α).*
3. *if $X \subset B' \subseteq B$ then $\alpha \in Cn(B')$ (and is maximal).*

A *selection function* is any function that "chooses" at least one element of $B \bot \alpha$. Formally:

Definition 3 (selection function). *An selection function γ for belief base B is a function such that for every α:*

1. $\emptyset \neq \gamma(B \bot \alpha) \subseteq B \bot \alpha$ *if* $B \bot \alpha \neq \emptyset$
2. $\gamma(B \bot \alpha) = \{B\}$ *otherwise.*

The *partial meet contraction* $-_\gamma$ is defined as follows: $B -_\gamma \alpha = \bigcap \gamma(B \bot \alpha)$.

The postulates that fully characterize partial meet contraction are the same that characterize kernel contraction with one exception. Core-retainment is replaced by a slightly stronger postulate called *relevance* [13]:

(relevance) If $\beta \in B \setminus B - \alpha$ then there is B' such that $B - \alpha \subseteq B' \subseteq B$, $\alpha \notin Cn(B')$ and $\alpha \in Cn(B' \cup \{\beta\})$

3   Architecture of the Proposed Framework

Belief revision theory is based upon a very well defined mathematical framework. This fact allows multiple operators to be compared theoretically, and a common conceptual base allows easier communication among members of the scientific community. This work aims to create a computational framework aligned with its mathematical foundations. Mathematical concepts and operators are translated directly to the computational framework, hence when a concept appears on several places it will be possible to reuse it computationally.

Although most of the work on belief revision in the literature has been proposed based on classical propositional logics, several results are independent of the underlying logic. Different logics and entailment operators can be used, provided that they meet some requirements, such as compactness [14, 15].

The proposed framework follows the same principles: every entailment operator must be usable through a common API, and users must depend only on this API, therefore entailment operators are interchangeable. Belief revision built depending on an abstract entailment operator can be used with any logic, sentence representation or entailment operator implementation. Interchangeability of this operator is central to the proposed framework, bringing to it a powerful abstraction used throughout all theoretical works.

In a deeper sense, the mathematical framework uses composition heavily to build increasingly complex concepts. This construct is directly related to object oriented programming concepts as composition and dependency injection: if a concept depends upon an other, it receives it upon construction.

This effort to keep an architecture aligned with its mathematical foundation goes beyond the entailment operator, all concepts have a well defined API: sets, sentences, revision operators, contraction operators, remainder set operators, selection functions, kernel operators and incision functions.

The benefits of this coherence are noticeable. Implementations are independent of logic, hence it is possible to plug in different entailment operators.

Writing new operators becomes a much easier task, since all basic concepts are already defined: translation from formulas and pseudo-code to real working code is direct. The framework comes with a well equipped unit testing structure, making it easy to write test cases that assert the algorithm's good behavior.

The framework comes with a powerful set of parameterizable tools capable of rendering thousands of pseudo-random scenarios. As all operators share a common API it is possible to measure the performance costs of several algorithms on the same scenarios. Since all operators share the base concepts such as entailment operators and knowledge representation, an implementation consists solely of its algorithm, therefore comparisons are fair.

4   Implementation Details

The proposed framework is written in Java. This language has features adequate to the proposal, having a fair performance, support for object orientation and widespread adoption that eases the learning curve for collaborators.

As described above, key elements of the framework are its abstractions. Some of them will be shown below (comments and imports omitted for brevity).

4.1   Sentence

```
public interface Sentence<S extends Sentence<S>> {
    S negate();
}
```

A sentence is anything that can return its negation. All sentences must be immutable, there are no other restrictions on the internal representation of the sentence or on the underlying logic.

Currently there are working implementations for propositional sentences and description logic axioms (OWL2[1]). Note that although negation is desirable it might be left unimplemented, limiting it's usage to algorithms that don't depend on the feature.

4.2   Immutable Set

```
public interface ISet<E> extends Iterable<E> {
ISet<E> union(ISet<E> elements);
ISet<E> union(E element);
ISet<E> minus(E element);
ISet<E> minus(ISet<?> elements);
ISet<E> intersection(ISet<?> elements);
boolean contains(Object element);
boolean containsAll(ISet<?> elements);
```

---

[1] http://www.w3.org/TR/owl2-overview/

```
boolean isEmpty();
int size();
}
```

The concept of set is ubiquitous on almost all mathematical domains, belief revision is no exception. Usually sets are considered immutable, and have well established operations ($\cup$, $\cap$, $\subset$, $\setminus$). Having an immutable set implementation with these usual operations is very useful when writing algorithms, rendering very expressive code.

For instance, the formula $A = B \cap (C \setminus D)$ can be implemented as follows:

```
A = B.intersection(C.minus(D))
```

Makings sets easy to manipulate grants easy translation from algorithm descriptions to working code, with lower chance for error and lower learning curve.

The immutable set implementation is vital to all algorithms, and must be very fast and have a very low memory footprint, since it is very common to have a huge number of such sets during a hard problem resolution.

4.3   SATReasoner

```
public interface SATReasoner<S extends Sentence<S>> {
    boolean isSatisfiable(ISet<S> sentences);
    boolean isUnsatisfiable(ISet<S> sentences);
    boolean entails(ISet<S> sentences, S sentence);
}
```

Although SAT is usually associated with propositional logic, this is simply a reasoner that can answer if a set of sentences is satisfiable (isSatisfiable). The two other methods (isUnsatisfiable and entails) can be directly described as a call to the first, but are added to promote clarity on client code, hence the following base abstract implementation:

```
public abstract class AbstractSATReasoner<S extends Sentence<S>>
    implements SATReasoner<S> {
    public boolean entails(ISet<S> sentences, S sentence) {
        return isUnsatisfiable(sentences.union(sentence.negate()));
    }
    public boolean isUnsatisfiable(ISet<S> sentences) {
        return !isSatisfiable(sentences);
    }
}
```

There are currently three implementations of SATReasoner, proving that reasoners are really interchangeable. It is possible to reason on propositional sentences using either SAT4J[2] or MiniSAT[3], and OWL axioms through HermiT[4].

The code that deals directly with the operator is naturally bound to an specific logic, sentence representation and with the reasoner's API. It constitutes a single point of contact with the reasoner, shielding all client code from specificities.

4.4   RevisionOperator

```
public interface RevisionOperator<S extends Sentence<S>> {
    ISet<S> revise(ISet<S> base, S alpha);
}
```

A revision operator acts upon a belief base and a sentence alpha, resulting in a consistent belief base that entails alpha.

An example implementation of the revision operator is an internal kernel revision operator. It is usually expressed by the following equation: $B \mp_\sigma \alpha = (B \setminus \sigma(B \perp\!\!\!\perp \neg\alpha)) \cup \{\alpha\}$. It's implementation is straightforward:

```
public class InternalKernelRevisionOperator<S extends Sentence<S>>
        implements RevisionOperator<S> {
    private final KernelOperator<S> kernelOperator;
    private final IncisionFunction<S> incisionFunction;
    public ISet<S> revise(ISet<S> base, S alpha) {
        return base.minus(incision(kernel(base, alpha.negate()))).union(alpha);
    }
}
```

The kernel operator and the incision function are also abstractions, and their implementations are also interchangeable. This separation allows a kernel operator implementation to focus on the algorithm, and lets an incision function focus on its heuristics.

Similarly, the revision operator implemented through partial meet is implemented as follows:

```
public ISet<S> revise(ISet<S> base, S alpha) {
    return intersection(this.selection
            (this.remainder(base, sentence.negate())))).union(sentence)
}
```

Notice that immutable data structures allow for a functional style that closely resembles the original formulae, thus implementing multiple operators that use the same base concepts is very simple.

---

[2] http://www.sat4j.org/
[3] http://minisat.se/
[4] http://www.hermit-reasoner.com/

5   Case Study: Kernel vs. Partial Meet Contraction

In this section, we illustrate the use of our framework by means of a case study. We want to compare kernel and partial meet base contraction operators with respect to time and calls to the SAT solver.

5.1   Theory

Through the logical characterization of both operations in terms of rationality postulates, we already know that kernel contraction is more general than partial meet contraction, i.e., for every partial meet operation it is possible to find an incision function and obtain the same operation using kernel contraction, but there are kernel contractions for which we cannot find a selection function giving the same result.

In order to compare the two kinds of operators, let us first consider the task of, given B and α, building $B\bot\alpha$ and $B \perp\!\!\!\perp \alpha$. As was shown in [16], given the remainder set it is possible to extract the kernel and vice-versa, using the idea of minimal cuts [17]. Each minimal cut of the kernel corresponds to (the complement of) a remainder and each cut of the complements of the remainders corresponds to a single kernel.

Some relations can also be established between incision and selection functions, as shown in [18]. First of all, from any selection function γ, an incision function σ_γ can be generated: $\sigma_\gamma(B \perp\!\!\!\perp \alpha) = B \setminus \gamma(B\bot\alpha)$.

Since not all kernel contractions are partial meet contractions, it is not always possible to derive a selection function from an incision function that will give the same results. For two special cases it is easy to see the correspondence: (i) [18] If an incision function is minimal, then it corresponds to a selection function that selects a single remainder; (ii) if an incision function is maximal, then it corresponds to a selection function selecting all remainders.

5.2   Practice

Even though it has already been shown that in theory the problem of finding the kernel or the remainder set can be reduced to each other, we want to find out whether one of them is generally more efficient than the other in practice.

We show here one set of tests that has been run using our framework. We do not claim that this answers the general question of which one should be used. Due to space limitation, in this paper we chose to focus on the testing framework.

Two kinds of algorithms were implemented, following what were called Blackbox and Glassbox approaches in [19]. The *BlackBox* algorithms are independent of the reasoner used, while the *Glassbox* algorithms use specific heuristics. The algorithms MaxSat and MUS are versions of the Glassbox approach used to find maximal satisfiable subsets and minimal unsatisfiable subsets respectively.

The idea here was to vary the number of clauses (denoted by m) and propositional variable letters (denoted by n) and measure the time used to find all the kernel sets or the remainders.

For each pair of values m, n, where m varies from 30 to 39 and n varies from 5 to 9, 20 different scenarios were generated randomly and each scenario was run 10 times for each of the four algorithms, yielding a total of 40,000 runs. The framework comes with several decorators that allow for measurement of all calls made to algorithms and reasoners, using the system's most precise clock.

The tests were run in a computer with a Core i7 870 processor, 2.93GHz running Linux and OpenJDK 1.6.0_23.

The tables below show the results obtained in milliseconds. We can see that for these particular scenarios, finding the remainder set is faster than finding the kernel. A more thorough comparison between kernel and partial meet operators falls out of the scope of these paper and is left for future work.

m\n	5	6	7	8	9
30	15.548	52.660	51.630	298.118	260.522
31	22.787	161.314	277.351	159.599	131.981
32	25.830	52.071	74.194	249.578	515.713
33	31.833	402.131	239.245	471.565	217.983
34	47.263	157.471	261.707	385.782	1506.652
35	49.971	147.430	541.548	1053.285	553.669
36	69.299	240.999	612.871	369.741	2763.529
37	68.659	263.568	405.763	626.081	887.284
38	130.399	365.039	681.860	1399.584	2434.254
39	144.728	498.095	818.494	1406.332	1119.880

(a) Kernel - Blackbox

m\n	5	6	7	8	9
30	6.553	15.721	15.262	73.385	131.277
31	10.848	41.050	80.825	61.200	24.808
32	11.019	19.640	22.751	101.171	288.677
33	10.657	85.891	44.641	328.649	44.518
34	15.088	41.297	58.147	94.284	433.727
35	16.174	53.854	110.325	268.098	310.166
36	17.391	64.899	100.688	153.783	189.082
37	19.515	46.233	240.970	295.847	525.697
38	38.461	61.911	349.425	305.350	871.313
39	29.720	129.286	312.918	622.218	405.000

(b) Kernel - MUS

m\n	5	6	7	8	9
30	13.230	40.791	49.918	194.875	301.245
31	18.216	65.908	132.554	122.561	110.405
32	19.029	31.759	57.391	421.925	654.748
33	19.186	178.126	176.347	309.462	293.210
34	27.554	66.210	118.724	272.623	1130.152
35	32.467	78.723	172.691	545.966	997.524
36	31.001	83.589	230.101	330.407	1273.092
37	33.432	105.958	270.637	484.591	765.010
38	42.084	134.350	510.763	689.396	1767.225
39	51.383	124.866	519.372	829.942	853.258

(c) Partial Meet - Blackbox

m\n	5	6	7	8	9
30	2.103	4.887	5.323	52.303	22.555
31	3.517	8.479	15.312	9.918	9.798
32	3.194	3.982	6.713	30.047	85.061
33	3.127	14.247	11.436	84.085	14.513
34	4.669	9.663	10.675	30.251	362.422
35	4.757	8.647	17.508	176.382	134.391
36	4.765	11.449	15.067	30.248	99.019
37	5.189	10.860	32.237	40.954	114.315
38	6.755	13.501	50.481	79.099	202.433
39	6.818	12.914	50.385	85.091	139.479

(d) Partial Meet - MaxSAT

6   Conclusions and Future Work

In this paper we have described a framework for empirically testing different belief change operators. The framework was implemented in Java and allows for the generation of testing scenarios and implementing different operators. We have shown an example where we compare the time consumed for finding kernels and remainder sets.

Future work includes using the framework and planning strategies for complete tests. We hope that these strategies will help us to create a benchmark for belief change, so that different systems and operators can be compared.

Another point that we would like to explore is the usability of the framework for different logics. More specifically, we want to focus on different fragments of classical propositional logics, such as the ones described in [20], and on different Description Logics, as described in [19, 21].

All source code is available at *http://code.google.com/p/bcontractor/*.

References

1. Gärdenfors, P.: Knowledge in Flux - Modeling the Dynamics of Epistemic States. MIT Press (1988)
2. Hansson, S.O.: A Textbook of Belief Dynamics. Kluwer Academic Press (1999)
3. Chou, T.S.C., Winslett, M.: The implementation of a model-based belief revision system. SIGART Bull. 2(3), 28–34 (1991)
4. Liberatore, P.: BReLS: a system for revising, updating, and merging knowledge bases. In: Proceedings of NRAC (1999)
5. Williams, M.A., Sims, A.: Saten: An object-oriented web-based revision and extraction engine. In: Proceedings of NMR 2000 (2000)
6. Alchourron, C.E., Gardenfors, P., Makinson, D.: On the Logic of Theory Change: Partial Meet Contraction and Revision Functions. The Journal of Symbolic Logic 50(2), 510–530 (1985)
7. Alchourrón, C., Makinson, D.: On The Logic of Theory Change: contraction functions and their associated revision functions. Theoria 48, 14–37 (1982)
8. Alchourron, C.E., Makinson, D.: On the logic of theory change: Safe contraction. Studia Logica 44(4), 405–422 (1985)
9. Grove, A.: Two Modellings for Theory Change. Journal of Philosophical Logic 17, 157–170 (1988)
10. Gärdenfors, P., Makinson, D.: Revision of Knowledge Systems Using Epistemic Entrenchment. In: Vardi, M. (ed.) Proceedings of TARK, pp. 83–95. Morgan Kaufmann (1988)
11. Hansson, S.O.: Reversing the {L}evi Identity. Journal of Philosophical Logic 22, 637–639 (1992)
12. Hansson, S.O.: Kernel Contraction. Journal of Symbolic Logic 59, 845–859 (1994)
13. Hansson, S.O.: A dyadic representation of belief. In: Gärdenfors, P. (ed.) Belief Revision. Cambridge Tracts in Theoretical Computer Science, vol. 29, pp. 89–121. Cambridge University Press (1992)
14. Hansson, S.O., Wassermann, R.: Local Change. Studia Logica 70(1), 49–76 (2002)
15. Flouris, G., Plexousakis, D., Antoniou, G.: Generalizing the AGM postulates: preliminary results and applications. In: Proceedings of NMR, pp. 171–179 (2004)
16. Ribeiro, M.M.: Belief Revision in Description Logics and Other Non-classical Logics. PhD thesis, University of São Paulo (2010)
17. Reiter, R.: A theory of diagnosis from first principles. Artificial Intelligence 32, 57–95 (1987)
18. Falappa, M.A., Fermé, E.L., Kern-Isberner, G.: On the Logic of Theory Change: Relations Between Incision and Selection Functions. In: Proceedings of ECAI, pp. 402–406. IOS Press (2006)
19. Kalyanpur, A.: Debugging and Repair of OWL Ontologies. PhD thesis, University of Maryland (2006)
20. Creignou, N., Papini, O., Pichler, R., Woltran, S.: Belief revision within fragments of propositional logic. In: Principles of Knowledge Representation and Reasoning (KR 2012). AAAI Press (2012)
21. Ribeiro, M.M., Wassermann, R.: Base Revision for Ontology Debugging. Journal of Logic and Computation 19(5), 721–743 (2009)

Sensorimotor Domain Approach for Artificial Autonomous Cognitive Development

Mauro E.S. Muñoz and Márcio Lobo Netto

Escola Politécnica da Universidade de São Paulo - POLI/USP
mesmunoz@usp.br, marcio.netto@poli.usp.br

Abstract. The autonomous cognitive development paradigm applied to cognitive agent's design assumes that no prior knowledge should be embedded while the agent's cognitive system is conceived. The use of this paradigm is an attempt to avoid the symbol grounding problem faced by some cognitive agent designs. We assume that the cognitive system of a sensorimotor agent is not aware of what it's sensors sense or what it's actuators do, or even where they are connected to in the agent's body. Using these prerogatives we present a *sensorimotor domain approach* that can be used to build agent's controls and to construct agents' cognitive systems. This proposal is then compared to an approach for sensorimotor control system implementation, here called *sensorimotor functional approach*, under the perspective of the autonomous cognitive development paradigm. In order to highlight the advantages of our proposal and to show how it can bypass the limitations of the functional approach, two simple examples are presented.

1 Introduction

The concept of *agent* in artificial cognitive science is used to emphasize the autonomy aspect of a theoretical model. The agent can thus be seen as a computational autonomous entity. This entity is supposed to interact in some environment. The agent perceives the environment through its sensors and can modify the environment through its actuators. The agent's sensors capture some environment aspects through the agent's body in the same way, the agent's body is used to modify the environment through the agent's actuators.

As an autonomous entity, the agent is designed to be responsible for some tasks. In cognitive science the agent is supposed to behave based on a cognitive mechanism. By cognitive agent we mean an agent able to build and handle symbolic representations of the environment. This knowledge may be organized in multiple layers giving the agent a better possibility to conduct its reasoning. But whatever this architecture may be, in an autonomous cognitive agent, it must follow an auto development process. Schemas, presented ahead are therefore a good basis to build a cognitive architecture. The cognitive mechanism is expected to improve the way the agent interacts with its environment, but also doing what its designer wants it to.

The approach used to design the agent's cognitive model determines the tasks the agent can ever accomplish. As the designer imposes on the agent its own

L.N. Barros et al. (Eds.): SBIA 2012, LNAI 7589, pp. 22–31, 2012.
© Springer-Verlag Berlin Heidelberg 2012

assumptions about the situations the agent can experience, the agent's cognitive system will be restricted to the designer's conception of the environment the agent is inserted in.

The Autonomous Cognitive Development (ACD) paradigm extends the autonomy of the agent's model to the process used to design the agent. Following this paradigm, the agent's cognitive mechanism should be capable to construct its own perspective of the environment it interacts to using the least possible knowledge from its designer. [1][1].

Piaget uses the *schema* concept to build a theory about the development of the intelligence in human beings. A schema is a cognitive entity capable to adapt itself using its interactions with its environment. It acts, perceives the results and modifies itself so that its structure increasingly reflects the relations between its actions and its sensations. This adaptation process grows extending the schema capabilities through time and space. [2], [3] and [4]. In Piaget's theory, schemata[2] are the basic cognitive units used to compose a schemata network. Schemata coordinate with other schemata to form the connected network at the first level. Also, schemata are the basis units used to compose a second and a third level of a hierarchical cognitive system that Piaget uses to explain the development of intelligence. [5].

Both the schema theory and the ACD paradigm assume the *agent* has no need to have prior knowledge about the environment it is inserted in. Also, the schema theory and agents in the ACD paradigm communicate with their environment exclusively through sensors and actuators. These prerogatives make possible the use of the schema theory to construct ACD-based agents. In fact, the use of the schema theory in the robotics and cognitive agents areas is not new. Arkin, in [6], proposes a hierarchical system using a motor schema as the basic unit of behavior specification for the navigation of a mobile robot. Drescher, in [7], proposes a hierarchical system based on the schemata theory where schemata are trained using statistical methods. Other schemata-based agents were proposed in [8], [9], [10], [11], [12] and [13]. Also, some schemata models were proposed in [14] and [15].

The use of the ACD paradigm to design autonomous agents not only fits the schemata theory, but it is a way to avoid the Symbol Grounding Problem (SGP). As stated by Harnad [17], the symbol grounding is the answer to the question: "How can the semantic interpretation of a formal symbol system be made intrinsic to the system, rather than just parasitic on the meanings in our heads? How can the meanings of the meaningless symbol tokens, manipulated solely on the basis of their (arbitrary) shapes, be grounded in anything but other meaningless symbols?"[3]. According to Taddeo and Floridi [18] the Symbol

---

[1] Weng uses the term Autonomous Mental Development, but we prefer to call it Autonomous Cognitive Development as *mental* term is not so used in the computational intelligence field.

[2] Plural of schema.

[3] Harnad uses the famous "Chinese Room Argument" of Searle [16] to introduce the SGP.

Grounding is the problem of how an artificial agent can "autonomously elaborate its own semantics" through interacting with its environment.

It should be noted that the SGP is not a problem only for agents that deal with symbolic processing, but is a challenge to other agent systems as in robotics. In [19], Williams argues that for agents like robots to respond appropriately to novel and unforeseen situations (situations for which they have not been explicitly programmed), they need sophisticated grounding capabilities. Novel situations need to be interpreted meaningfully by the agent - and not just by the designer a priori - but in real time.

Taddeo and Floridi, in [17], establishes a *zero semantic commitment condition* (ZSCC) that must be satisfied by systems attempting to solve the SGP. This commitment states (1) no form of *innatism* is allowed; no semantic resources should be presupposed as already pre-installed in the agent; and (2) no form of *externalism* is allowed either; no semantic resources should be uploaded from the "outside" by some *deus ex machina* already semantically-proficient.

In this article we present a basic approach for a schema model that can be used for autonomous agent design that follows the ACD paradigm and that respects the *zero semantic commitment condition*.

2   Motivation

A sensorimotor schema is a cognitive mechanism that can act in certain circumstances to achieve its own intrinsic goals. Examples of what Piaget calls schemata are the *grasp* schema, the *suction* schema, the *look-at* schema, etc. Thus, a schema can keep its execution for a certain period of time. In the same way it can use different sensory stimuli during its execution to achieve the goal.

We will focus in just one moment of time (a time tick) in the schema execution herein to try to characterize the properties the schema should have to fit in ACD and ZSCC prerogatives. To do so, first we will describe the agent and schema properties.

The agent has sensors and actuators to interact with its environment. Let $\bar{S}$ be the set of all agent sensor states and $\bar{A}$ be the set of all agent actuator states. In the same way the sensorimotor schema has its own sensors and actuators which are taken from the agent's sensorimotor apparatus. Let $S \subseteq \bar{S}$ be the set of all schema sensor states and $A \subseteq \bar{A}$ be the set of all schema actuator states. We will call *schema sensation*, or just *sensation* the schema sensors state. In the same way we will call *schema action*, or just *action* the schema actuators state. The schema sensation at time $t \in \mathbb{N}$ will be given by $s_t \in S$ and the schema action will be given by $a_t \in A$.

When the schema is in execution and its sensation is given by s_t, the schema is expected to be able to suggest some action a_t and also preview the next sensors state s_{t+1}. We will represent this relation by:

$$s_t \xrightarrow{a_t} s_{t+1} \tag{1}$$

The relation (1) expresses the schema *atomic causality behaviour* for the schema execution in the sense that it is the thinnest granularity in time the schema's prediction can reach.

Let us now see how some of the existent schema models are implemented.

Drescher [7] uses the idea of *items* to model his agent sensors and actuators. The items used as Boolean values indicating the presence or absence of a token. Each schema is created with an action. The application of this action is used to find the correlations between the schema context (sensors) and its results (prediction). Holmes [9] and Chaput [11] also use the idea of *item* and associate one single action with each schema. In [11] the schema is implemented by a self organized map (SOM).

Tani [13] uses a parameterized recurrent neural network (RNNPB) to generate sensors and actuators patterns. Sensors (s_t) and actuators (a_t) states are inserted in the RNNPB and the next sensor (s_{t+1}) and actuator (a_{t+1}) states are taken from the RNNPB output and reused as the RNNPB input for the next step. The schema training and usage is defined by three phases: (1) learning phase: the parametric biases (PB) and the matrix weight values are found using the training data. This is the *forward* in which the recurrent network link is not connected; (2) generation phase: fixes PB to find actions and predicted sensations. Now the recursion link is connected. (3) recognition phase: when some pattern is given the *way to act* is found by using an iterative inverse computation. Glaser [14] uses Tani's idea, but uses a schemata map activity to act as Tani's PB. It is important to note that Tani's and's Glaser models use the idea of *forward* and *inverse* functions. The forward function is used to find a possible action based on a sensation, and the inverse function is used to find a sensation after an action is applied.

One interesting thing to note in all schema models we studied is that they all assume the actuators state is a function of the sensors state, even at the schema atomic causality behaviour context. We will call this assumption the *sensorimotor functional approach*. In other words, using (1), this assumption states that:

$$\exists f \mid a_t = f(s_t), \ f : S \mapsto A \qquad (2)$$

In the next section we will show that this assumption does not always capture some essential relations between sensations and actions of a schema when the ACD paradigm is adopted together with the ZSCC prerogatives.

3   Sensorimotor Domain Approach

The causality atomic behaviour expresses the core problem faced by schema model designers. The question: how to find which actions can be taken in what sensor context to achieve what sensor result? The answer for this question reflects the schema model conception the designer will follow.

As the agents start without any knowledge about the environment they are inserted in, it is usual to assume they act randomly. This random behaviour generates the sensorimotor data that will be used to find sufficient corelations to consolidate the schema structure. The structure should be general enough to hold all corelations the sensorimotor schema can present.

As we assume actions are not functions of sensations, another way to tie them together is needed. We propose that the sensor and motor space should be used in an indistinguishable way by the schema structure. In our approach we tie sensations and actions using classes:

$$[s|a] \in C \Leftrightarrow s \xrightarrow{a} p \wedge p \in G \tag{3}$$

Where: $s \in S$ is the current sensors state; $a \in A$ is the current actuators state; $C \subseteq S \times A$ is the class of usefull situations; $p \in \bar{S}$ is an agent's sensation prediction (or expected event); and $G \subseteq \bar{S}$ is the set of all schema's goals.

A sensation $s \in S$, and an action $a \in A$, perceived by the schema are concatenated to produce what we will call a sensorimotor situation: $x = [s|a] \in S \times A$. The sensorimotor situation data x are then used to delineate the C class. A mandatory question is: what does this class classify? It classifies the achievement of something important for the schema, i.e. the schema goals. These goals will be expressed by a set of the agent's sensors states represented by G. Note that G is a subset of the *agent's* sensations ($\bar{S}$) and it is not restricted to the *schema* sensations (S).

The first consideration we will make to eliminate the use of the designer's knowledge from the schema model conception will be to treat sensations and actions as signals, i.e. $\bar{S} \subset \mathbb{R}^n$ and $\bar{A} \subset \mathbb{R}^m$.

The second consideration is that the agent's designer should not know the contribution of each agent sensor or actuator to the agent's tasks. In the same way, the schema designer does not know about the connections between the schema and the agent sensors or actuators.

Now let us see two cases which the functional approach does not suffice to represent the schema atomic behaviour expressed by (1).

Multiple Actions Generating the Same Result: In this case, when the schema sensors state is s_t, the resulting sensation s_{t+1} can be achieved using any action $a \in A_p \subseteq A$.

To ilustrate this case imagine that the predicted (expected) schema sensation s_{t+1} represents the perception of two fingers touching themselves. The schema actions are translated to the fingers' movements and the schema sensations are read from the fingers' relative positions. Hence, when the fingers are very close, in the next time step it is clear that there are more than one way to put they together.

Using the functional approach an action solution can be given by $a = f(s_t)$, but note that every time the schema faces sensation s_t the same action $f(s_t)$ will be imposed by the schema application. The agent will be predestined to always touch its fingers the same way.

On the other hand, using the proposed domain approach, the set of all possible sensorimotor situations that leads to both fingers touching will be stored in the C class, and can be represented by $A_p = \{a \in A \mid [s_t|a] \in C\}$.

In the case presented here one can argue that if set A_p of possible solutions is a connected set then the functional approach can give a viable solution. But clearly this is not true when A_p is a disconnected set. In section 4 we implement an example of this case and show that the functional approach is not capable to solve the problem.

One point to note in case A_p is disconnected is that the designer can use more than one schema to handle the situation, one for each connected subset of A_p. But when the designer does this, it is using its own knowledge about the agent's interaction with the environment to conceive the agent's model, which goes against our prerogatives.

Irrelevant Actuators: This is the case in which some of the schema actuators are unimportant to the actions necessary to achieve the expected schema goal.

As a concrete example, let us say the schema has two sets of sensors, S^L and S^R connected to the left and right hands of the agent, respectively ($s \in S^L \times S^R$). In the same way the agent has two sets of actuators, A^L and A^R also connected to left and right hands ($a \in A^L \times A^R$). Now suppose the schema goal is to touch an object with its right hand. The object is in a fixed position with respect to the agent's hands. Now, let $s_r \in S^R$ and $a_r \in A^R$ represent a right hand sensation and action leading to the schema's goal in the next time step. It is worth observing that the designer can not use the knowledge of this scenario to conceive the schema.

Using the functional approach we have:

$$[a_l | a_r] = f([s_l | s_r]) \qquad (4)$$

for any left hand sensations $s_l \in S^L$ and actions $a_l \in A^L$. Note that any time the schema sensations are given by $[s_l | s_r]$ the left hand will be moved by a_l, when in fact it could be doing any other thing. In other words the left hand movement will be tied to the right hand movement even when it has no role to play in the goal achievement. So, both are bound together by some artifact that imposes some relation between both actions, which might not exist (they could behave independently).

Using the domain approach, we have:

$$[s_l | s_r | a_l | a_r] \in C \Leftrightarrow [s_l | s_r] \xrightarrow{[a_l | a_r]} p \wedge p \in G \qquad (5)$$

And the schema action suggestion set will be given by:

$$A_p = \{ [a_l | a_r] \in A \mid [s_l | s_r | a_l | a_r] \in C \} \qquad (6)$$

As we assume the left and right hands actuators and sensor do not have crossed influence, we know that all sensorimotor situations $x = [s_l | s_r | a_l | a_r]$ that lead to the schema goal will be captured by the schema C class. In other words, it does not matter what values variables s_l and a_l assume, all points with s_r and a_r that lead to the goal will be included in C. The schema suggestion set can be expressed by:

$$A_p = [a_l | a_r] \in A \mid [s_l | s_r | a_l | a_r] \in C, \ \forall s_l \in S^L \wedge \forall a_l \in A^L \qquad (7)$$

4   Implementation

In this section we show the practical use of the domain approach to solve a simple artificial problem created to highlight the difference between the functional and domain approaches.

The Problem: The agent is at a fixed position and it should throw a ball to hit a basket located somewhere ahead, as shown in figure 1. Note that the basket is composed by two disconnected mini-baskets, but the agent does not know about this. The agent always throws the ball at the same fixed angle. The agent has two sensors s and g and one actuator a, where $s, g, a \in \mathbb{B} = [-1, +1] \subset \mathbb{R}$.

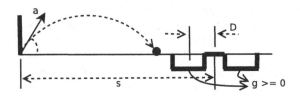

Fig. 1. Ilustration problem scenario

The schema is only connected to agent sensor s and to actuator a. Sensor s informs the distance between the agent and the center of the basket. Sensor g indicates if the ball hit the basket or not. Actuator a indicates the ball initial velocity.

The distance between the agent and the basket is normalized to the range $[-1, +1]$, where -1 means distance 0 (zero) and $+1$ indicates the maximum distance. In the same way the actuator value is normalized in the range $[-1, +1]$, where -1 represents velocity 0 (zero) and $+1$ represents the velocity needed to throw the ball at the maximum distance. Sensor g is set to $+1$ if the ball hit the basket and -1 otherwise.

Note that the distance the ball reaches is proportional to its initial velocity. As the distance sensor and the actuator are normalized to the same range, if the agent throws the ball using equation $a = (s - D)$, then the ball will hit the center of the first mini-basket, and if it throws it using $a = (s + D)$ the ball will hit the center of the second.

In both approaches, functional and domain, the agent is composed by just one schema trained with the same algorithm, described as:

```
1. For each time step, do:
    1.1 Put the basket at some random location;
    1.2 Calculate 's' based on the basket location;
    1.3 The schema generates a random action 'a';
    1.4 The environment takes the action and verifies if the ball hit the basket.
    1.5 'g' is set with +1 if the ball hit the basket or -1 otherwise.
    1.6 The schema is trained using sensor 's' and actuator 'a'.
```

Both schema implementations use only s and a for data training. Agent sensor g is reserved to indicate when the schema goal is achieved. The approaches were implemented using feed-forward neural networks (FFNNs), as ilustrated in figure 2.

Fig. 2. Sensorimotor functional **(a)** and domain **(b)** generic implementations using feed-forward networks. Both implementations use n sensors and m actuators.

Functional Approach Implementation: The mathematical model used to implement the functional approach is given by:

$$a = f(s) \text{ if } g \geq 0, \ f : \mathbb{B} \mapsto \mathbb{B}$$

The schema was implemented with a feed-forward neural network (FFNN) with 1 input (sensor s), 6 units in the hidden layer and 1 output (actuator a). The FFNN was trained using a simple backpropagation with gradient descent algorithm with learning rate of 0.1. If $g \geq 0$ then the FFNN is trained with s as input an a as output. If $g < 0$ then s and a values are ignored.

At each time step the last 24 training points are used by the learning algorithm. This strategy is used to speed up the neural network convergence.

Domain Approach Implementation: The mathematical model used to implement the domain approach is given by:

$$C(s, a) = \begin{cases} -1 & \text{if } g < 0 \\ +1 & \text{if } g \geq 0 \end{cases}$$

The schema was implemented with the same FFNN code and trained using the same simple backpropagation with gradient descent algorithm with learning rate of 0.1. The FFNN has 2 inputs, (s, a), 6 units in the hidden layer and 1 output (class indicator).

As was done for the functional approach, the last 24 positive and negative training points are kept and used in each training step.

Results: Figure 3 shows 3 experiment trails each one plotted after 5000 time steps. To guarantee that the functional and the domain schemata receive the same data, only one agent was used. This agent had both schemata working in parallel.

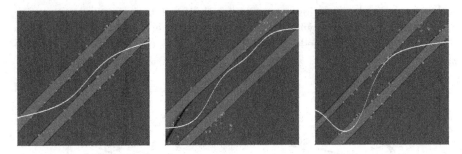

Fig. 3. The figure shows the schema mapping results for the functional (white line) and the domain (dark and grey) approach for 3 different runs of 5000 time steps each. The x-axis represents the values of sensor s in the range $[-1, +1]$ and the y-axis represents the actuator a in the same range.

The solution for the proposed problem are two parallel bands with centers in $a(s) = (s - D)$ and $a(s) = (s + D)$. Note that for all runs the domain approach is near the solution. For this approach, the grey area are the points belonging to class C and the dark one are the points not in C. For the functional approach, the white curve shows the points of the function f $(a = f(s))$. As one can note, in each run the function expresses a different solution and few points belong to the solution bands and it does not converge to an acceptable solution.

5   Conclusion and Future Work

In this work, we characterized the functional approach used to build cognitive agent based on the schema theory. We showed that when fitting the schema theory with the autonomous cognitive development and the zero semantic commitment, the functional approach is not enough for some situations we pointed out. We then proposed a sensorimotor domain approach that can handle those situations. The next step of our research is to study how schemata using the sensorimotor domain approach can be combined to build a complete cognitive system.

References

1. Weng, J., McClelland, J.L., Pentland, A.: Autonomous Mental Development by robots and animals. Science 291(5504), 599–600 (2001)
2. Piaget, J.: O Nascimento da Inteligência na Criança. Publicações Dom Quixote, Lisboa 1986. Original: La Naissance de l'intelligence chez l'enfant. Delachaux & Niestlé S.A (1977)
3. Piaget, J.: A Construção do Real na Criança. Editora Ática, 3a ed, 1996. Original: La Construction du Réel chez l'enfant. Delachaux & Niestlé S.A., 3a ed (1977)
4. Piaget, J.: A Formação do Símbolo na Criança: imitação, jogo e sonho, imagem e representação. Editora LTC, 3a ed, 1990. Original: La Formation du Symbole chez L'enfant: Imitation, Jeu et Rêve, Image et Représentation. Delachaux & Niestlé S.A., 3a ed (1964)

5. Piaget, J.: O Desenvolvimento do Pensamento - Equilibração das Estruturas Cognitivas. Publicações Dom Quixote, 1a ed, 1977. Original: L'Equilibration des Structures Cognitives. Presses Universitaires de France, 1a ed (1977)
6. Arkin, R.C.: Motor Schema Based Navigation For A Mobile Robot: An Approach To Programming By Behavior. In: Proc. ICRA (1987)
7. Drescher, G.L.: Made-Up Minds: A Constructivist Approach to Artificial Intelligence. The MIT Press (1991)
8. Yavuz, A., Davenport, D.: PAL: A Constructivist Model of Cognitive Activity. In: Proc. Int. Conf. on New Trends in Cognitive Science: Does Representation Need Reality? Viena, Autria (1997)
9. Holmes, M., Isbell, C.: Schema Learning: Experience-based Construction of Predictive Action Models. In: Advances in Neural Information Processing Systems, vol. 17 (2005)
10. Wazlawick, R.: Um Modelo Operatório para Construção de Conhecimento. Tese (doutorado) - Universidade Federal de Santa Catarina, Florianópolis, BR - SC (1993)
11. Chaput, H., Kuipers, B., Miikkulainen, R.: Constructivist Learning: A Neural Implementation of the Schema Mechanism. In: Proceedings of the Workshop on Self-Organizing Maps, WSOM 2003 (2003)
12. Wolpert, D., Kawato, M.: Multiple paired forward and inverse models for motor control. Neural Netw. 11(7-8), 1317–1329 (1998)
13. Tani, J., Ito, M., Sugita, Y.: Self-organization of distributedly represented multiple behavior schemata in a mirror system: reviews of robot experiments using RNNPB. Neural Netw. 17(8-9), 1273–1289 (2004)
14. Claudis, G., Joublin, F., Goerick, C.: Learning and Use of Sensorimotor Schemata Maps. In: IEEE 8th International Conference on Development and Learning (2009) Complemento
15. Muñoz, M.E.S.: Proposta de um Modelo de Esquema Cognitivo Sensório-Motor Inspirado na Teoria de Jean Piaget. Dissertação (mestrado) - Universidade Federal do Rio Grande do Sul, Porto Alegre, BR - RS (1999)
16. Searle, J.R.: Minds, brains, and programs. Behavioral and Brain Sciences 3, 417–424 (1980)
17. Harnad, S.: The Symbol Grounding Problem. Physica D: Nonlinear Phenomena 42, 335–346 (1990)
18. Taddeo, M., Floridi, L.: Solving the Symbol Grounding Problem: A Critical Review of Fifteen Years of Research. Journal of Experimental and Theoretical Artificial Intelligence 17, 419–445 (2005)
19. Williams, M.-A., McCarthy, J., Gárdenfors, P., Stanton, C., Karol, A.: A Grounding Framework. Autonomous Agents and Multi-Agent Systems 19(3), 272–296 (2009)

A Service-Oriented Architecture
for Assisting the Authoring of Semantic Crowd Maps

Henrique Santos and Vasco Furtado

Universidade de Fortaleza - UNIFOR, Fortaleza CE 60.811-905, Brazil
PPGIA – Programa de Pós-Graduação em Informática Aplicada
hensantos@gmail.com, vasco@unifor.br

Abstract. Although there are increasingly more initiatives for the generation of semantic knowledge based on user participation, there is still a shortage of platforms for regular users to create applications on which semantic data can be exploited and generated automatically. We propose an architecture, called Semantic Maps (*SeMaps*), for assisting the authoring and hosting of applications in which the maps combine the aggregation of a Geographic Information System and crowd-generated content (called here crowd maps). In these systems, the digital map works as a blackboard for accommodating stories told by people about events they want to share with others typically participating in their social networks. *SeMaps* offers an environment for the creation and maintenance of sites based on crowd maps with the possibility for the user to characterize semantically that which s/he intends to mark on the map. The designer of a crowd map, by informing a linguistic expression that designates what has to be marked on the maps, is guided in a process that aims to associate a concept from a common-sense base to this linguistic expression. Thus, the crowd maps start to have dominion over common-sense inferential relations that define the meaning of the marker, and are able to make inferences about the network of linked data. This makes it possible to generate maps that have the power to perform inferences and access external sources (such as DBpedia) that constitute information that is useful and appropriate to the context of the map. In this paper we describe the architecture of *SeMaps* and how it was applied in a crowd map authoring tool.

1 Introduction

Motivated by the huge success of Wikipedia, wiki applications have not been restricted to crowdsourcing via text sharing. On the contrary, there has recently been an explosion of interest in using the Web to create, assemble, and disseminate geographic information provided voluntarily by individuals. Crowd mapping activity, combining the aggregation of a Geographic Information System (maps on the Web) and crowd-generated content, flourishes daily [1, 2]. Sites such as *Wikimapia* (http://www.wikimapia.com), *Wiki-Crimes* (http://www.wikicrimes.org) [3], Click2fix (http://www.click2fix.co.sa), *Crowd-map* (www.crowdmap.com), and *OpenStreetMap* (http://www.openstreetmap.org) are empowering citizens to create a global patchwork of geographic information, while Google Earth and other virtual globes are encouraging volunteers to develop interesting applications using their own data. In crowd map applications, the digital map works as a

L.N. Barros et al. (Eds.): SBIA 2012, LNAI 7589, pp. 32–41, 2012.

blackboard for accommodating stories told by people about events they want to share with others typically participating in their social networks.

On the other hand, although there are increasingly more initiatives for the generation of semantic knowledge based on user participation (e.g.: twine, semantic wiki [4], and ontowiki [5]), there is still a shortage of platforms for the development of applications by non-expert users on which semantic data can be generated automatically and exploited by these applications.

Our work fits innovatively into this context. We propose a service-oriented architecture, called *SeMaps* (from Semantic Maps), for expressing the semantics of what the designer of the map intends to mark on the map (here called markers). We call "semantic characterization" the act of describing the concept(s) that best represent(s) a marker. The characterized concepts are associated with the linked data represented in RDF. In doing so, *SeMaps* makes it possible to generate Semantic Crowd Maps that have the power to perform inferences and/or access external sources that constitute useful and appropriate information to the map context. Wikipedia itself – through its RDF representation in DBpedia [6] – can be one of these external sources that provide additional and contextual information within the map.

In this paper we describe the architecture of *SeMaps*, with an emphasis on the specification of its own ontologies as well as its connection with Linked Open Data (LOD). Then we describe how the *SeMaps* services were coupled to a crowd mapping authoring tool in order to drive the marker's concept elicitation from the common-sense base in order to define the semantic value of what the designer wishes to mark on the map. An example of a crowd map generated from this approach demonstrates how easy the process of attributing semantics and accessing LOD in crowd maps can be.

2 Providing Semantics to Crowd Maps

At the core of *SeMaps* are the ontologies that describe the knowledge behind the maps created. Due to the collaboration characteristic inherent in the maps, we sought to reuse concepts of ontology that supported the collaborative model of content creation. We found the support we needed in the Semantically-Interlinked Online Communities (SIOC) ontology [7]. In addition to describing the content created by the participants, it also provides the possibility of adding semantics to such content, an important part of our model. In the *SeMaps* ontology, the marker has information about an event, a person, a business or a particular fact (represented by the concept classes created during the process of marker creation) and on the marker per se (such as date and time, the user who created it, among others). Moreover, in the marker there is a set of features to describe the provenance of the information, such as the type of source and how reliable it is. For representing these latter features, we imported the PML2 (Proof Markup Language) ontology [8]. However, we specialized some of the concepts to cope with the particular features of *SeMaps*, particularly for representing the notion of reputation inspired in [9] and [10]. As shown in Figure 1, the main classes of *SeMaps* are:

- Marker – the class responsible for describing a user's report about something of the domain of the map at issue. The property *sioc:has_creator* links to the user account who reports and the property *sioc:topic* links to the concept being reported.

- WikiUser/WikiUser Account – WikiUser describes a user registered on a particular crowd map. Those users make up a social network through the concepts of person and friends imported from the FOAF (friend of a friend) ontology [11]. WikiUser Account describes the account the user utilizes to interact with the map.
- Concept classes – describes the thing being reported on the crowd map. Instances of this class also link to an *InferenceNet* concept via the *rdf:type* property. As *SeMaps* assumes that a marker must be a person, an organization, an event, a complaint, an artistic production (movies, photos, books, etc.), a building, or a commercial establishment, every created class is a subclass of these classes (we dub them the *SeMaps* top classes).

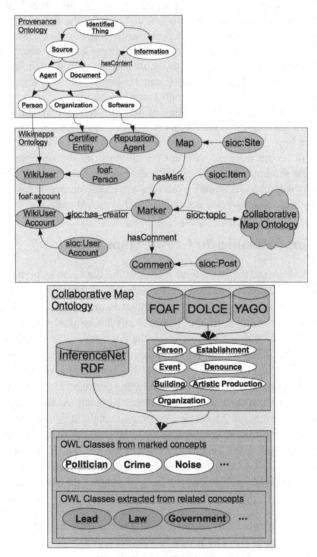

Fig. 1. SeMaps Ontologies

SeMaps allows the generated maps to have completely heterogeneous domains. Therefore, there is a need to describe types of information that exist only for a specific map. A particular map can refer to crimes, noise pollution, humanitarian assistance, etc. Each map contains specific information that must be described. To enable this specialized description, OWL classes are created on demand, i.e., *SeMaps* representation evolves according to the map created. These classes are extracted directly from the markers created on the various crowd maps and are images of the *sioc:topic* property. As previously mentioned, the instances of these classes are linked to a concept through the *rdf:type* property.

Besides using the aforementioned ontologies, *SeMaps* services rely on a bilingual knowledge base, which expresses inferentialist semantic and common-sense knowledge – *InferenceNet* [12]. *InferenceNet* expresses semantic content through a network that connects a concept to many others through dozens of common-sense semantic relations and that are inferential in nature (preconditions and postconditions of the use of concepts). Formally, this base is represented in a directed graph $G_c(C,Rc)$. Each inferential relationship $rc_j \in Rc$ is represented by a tuple (*relationName*, c_i, c_k, *type*), where *relationName* is the name of an *InferenceNet* semantic relation (Capableof, PropertyOf, EffectOf, etc.), c_i and $c_k \in C$ are concepts of a natural language, and type = "Pre" or "Pos", indicating a precondition or a postcondition for using the concept c_i.

InferenceNet is linked to YAGO and DBpedia [13], which allows rich inferences to be made, since the base expresses common-sense knowledge and the inferential import of the concepts in the reasoning.

3   The *SeMaps* Architecture

SeMaps is a set of web services (WS) specifications, also including built-in functionalities, capable of enabling semantics on crowd maps. The main features are: description of crowd maps in an RDF/OWL ontology, web services for semantically characterizing a maker, and web service for LOD contextualized resources retrieval from data already on the map. Figure 2 shows the *SeMaps* architecture and its relation with a crowd map.

3.1   Semantic Characterization Module

The module for semantic characterization comprises the services by which the user-designer can define the semantic value s/he wishes to mark on the map (the marker) based on the common-sense base. In short, the user-designer, by informing a linguistic expression that designates its marker, is guided in a process that aims to associate a concept from the common-sense base to the linguistic expression that names the marker. Thus, the map starts to have dominion over common-sense inferential relations that define the meaning of the marker and are able to make inferences about the network of linked data. *SeMaps* provides services to associate a concept to a marker and to create an ontology for the crowd map.

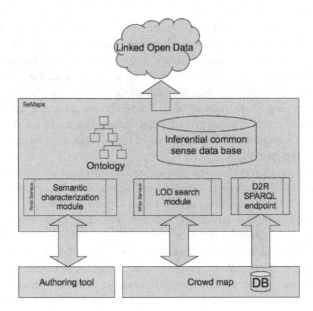

Fig. 2. SeMaps Architecture

The association of a concept to a marker is an iterative process that occurs while there are markers to be created by the designer. Figure 3 describes the interface of the WS (*SemanticCharacterization*) with its methods and I/O parameters. The *semantic-Characterize* method receives as a parameter the string that describes a marker and returns a list of URIs that identify resources found in the common-sense base represented in RDF.

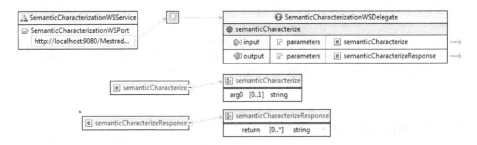

Fig. 3. Semantic characterization web service interface

For each marker with an associated concept, *SeMaps* uses another service to create the classes of the ontology for this map. The instances of the classes are linked to DBpedia/YAGO. Figure 4 describes the interface of the Concepts WS. The input parameter of the *createConcept* method is the *string* that describes the marker and returns a unique identifier for the concept. This number is the main reference of the concept in *SeMaps*.

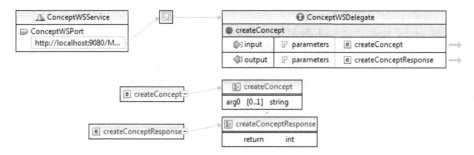

Fig. 4. Concept creation module web service interface

3.2   LOD Search Module

The search module of *SeMaps* provides a service that obtains resources (articles, news, etc.) from LOD. This search is based on the semantic characterization done in the module described in 3.1. The *searchLODForConcept* method of the LOD WS receives a concept representing a marker as input and returns a list of URIs describing resources related to the concept. Usually the algorithms that implement this search also use *InferenceNet*. An example of this was described in [14]. The URIs returned by the method can be used in different ways, for instance, used as a widget or plotted directly on the map. Due to space constraints, the web service description in WSDL was omitted.

3.3   D2R Endpoint

SeMaps assumes the use of a service of mapping relational data to RDF. This is necessary not only to avoid data replication, since typically the authoring tools have their data represented in RDBs, but also because the crowd map data will be available to be accessed by a SPARQL endpoint. We used the D2R server [15], which is an open and free system for publishing relational data on the Web. It enables RDF and HTML browsers to navigate the content of non-RDF databases, and allows applications to query a database using the SPARQL query language over the SPARQL protocol.

4   Integrating *SeMaps* into **WikiMapps**

The *SeMaps* architecture allows several different implementations for each service. For example, different algorithms for semantic annotation can be used in the process of semantic characterization. The only requirement that must be satisfied is to follow the interface specified previously.

   In order to put *SeMaps* to work, we integrated the *SeMaps* services for semantic characterization of a marker into *Wikimapps* (www.wikimapps.com), an authoring tool for creating and hosting crowd maps. The semantic characterization process and the search module were instantiated with algorithms described in [14].

We also used a D2R Server to maintain the map ontology. Thus, for each marker created in the authoring tool, *SeMaps* creates an entry in this table and a mapping in D2RQ helps to automatically update the ontology.

To exemplify how the knowledge represented on a crowd map generated from *Wikimapps* works, let's assume that a designer wants to create a map of politicians of a certain region. In order to associate semantics to the markers, *Wikimapps* was prepared to call the services for semantic characterization and concept creation of *SeMaps*. Thus, when the designer of a crowd map in *Wikimapps* defines a marker called "politician", it is automatically associated with *InferenceNet*'s "politician" concept that, in turn, is linked to instances of YAGO and DBpedia. The way each service of this process was implemented is described in [14].

The crowd map generated by *Wikimapps* was also prepared to use a *SeMaps* search WS. This service was called inside a widget that brings LOD data from the New York Times and DBpedia, with additional information related to subjects to which the markers of the map refer. The widget calls the WS whenever there is a change in the map's viewport.

Figure 5 presents the results of a query made by the crowd map, whereby news from the New York Times on corruption, scandals and laws has been found. Note that this information refers to the region the map selected (in this case the state of Illinois).

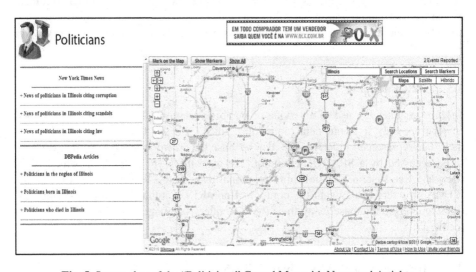

Fig. 5. Screen shot of the "Politicians" Crowd Map with News and Articles

Figure 5 also shows links representing the relations of common sense that have been identified. If the user decides to navigate the map, other queries are performed, this time filtering information regarding relationships as news stories on laws proposed by politicians, corruption, or any scandals. In the example shown in Figure 5, the possible issues were created by the inferential content of the concept "politician", expressed in *InferenceNet*: "a politician is able to propose laws" and "a politician has the capacity to corrupt." This common-sense knowledge is expressed in *InferenceNet* through the relationships (CapableOf, "politician", "to propose laws", "Pre") and (ProprietyOf, "politician", "corruption", "Pre"), respectively. Moreover, the concepts "political", "law" and "corruption" are associated – via LOD – with YAGO resources and DBpedia. Also in Figure 5, we see another characteristic obtained by the fact of knowing the semantics of the concept: DBpedia articles are related to the location indicated on the map. Further information regarding the place of birth, death, work and other references that are based on *GeoNames* allow *WikiMapps* to automatically sort the items for these characteristics.

5   Related Work

The production of linked open data is a growing trend on the web. Thus, the number of tools for generating semantic content on the Web is increasing. Highlighted among these tools are those that export pre-existing data without semantics, such as Drupal [16].

LinkedGeoData [17] is a project aimed at processing and representation of RDF data that was created collaboratively in *OpenStreetMap* (OSM – www.openstreetmap.org). In addition, the project aims to allow binding of OSM with other bases of the LOD. *LinkedGeoData* developed an ontology partly derived from the relational schema coming from *OpenStreetMap* and WGS84 (World Geodetic System). To map the collected data in instances of the OSM ontology, *LinkedGeoData* makes use of Triplify, a tool able to accomplish mapping from relational databases to RDF. Despite the similarities with *SeMaps*, *LinkedGeoData* is focused on the geographic elements that compose the map instead of modeling denounces, events, persons and other entities that are associated to places. This is not a disadvantage of *LinkedGeoData*, but shows how it is focused on collaborative cartography and it is not able to cover what *SeMaps* covers, like events or complaints, for instance.

Another group of tools that produce semantic content is that of semantic wikis. The semantic wikis, such as *Semantic MediaWiki* [4], allow the users themselves to describe – through notes – the content edited by them. Although they allow semantic data to be generated collaboratively, semantic wikis tend to make the process of creation somewhat difficult for users, since they need to use special tags in order for the data to be linked semantically. This difficulty can result in data linked erroneously or without sufficient links. Some projects such as Wolfram Alpha (www.wolframalpha.com) aim to provide a "computational knowledge engine" and allow the generation of knowledge widgets that are very similar to the services that can be produced using the services provided by *SeMaps*. However, a proprietary knowledge base is used, exploring alternative representational ways that could eventually support or complement our approach.

Hermes [18] is a system capable of creating a knowledge base and from there, filtering news from RSS feeds in accordance with the desired concepts. It compiles an ontology with linguistic expressions that appear in the selected news items. By creating a knowledge base on top of news, the proposed framework does not use a knowledge base of common sense as we do in *SeMaps*.

6   Conclusion

In this paper we propose the semantic crowd map concept, in which the process of participation and interaction always occurs in relation to a given space indicated on a digital map and the semantic of the marker is provided by services of *SeMaps*. We describe a platform for creating semantic crowd maps, which provides an environment for the creation and maintenance of sites based on crowd maps. We seek to supply the lack of platforms for the development of applications, by non-specialist users, where semantic data are automatically generated and exploited by these applications.

SeMaps allows a user who wishes to create a crowd map to do so with the description of the concept of that which s/he wishes to mark on the map. Hence, the maps produced by *SeMaps* have the power to make inferences and access external sources that constitute useful information, appropriate to the context of the map. This semantic characterization of the marker used a common-sense base in Portuguese and English. The integration of *SeMaps* into *Wikimapps* showed that crowd maps generated by *WikiMapps* are more informative, since before *SeMaps* they only were able to show what people marked on the map and nothing more. With *SeMaps*, they are able to access LOD and show relevant and contextualized information.

References

1. Mac Gillavry, E.: Collaborative Mapping and GIS: An Alternative Geographic Information Framework. Collaborative Geographic Information Systems, 103–119 (2006)
2. Rouse, L.J., Bergeron, S.J., Harris, T.M.: Participating in the geospatial web: collaborative mapping, social networks and participatory GIS. The Geospatial Web, 153–158 (2007)
3. Furtado, V., Ayres, L., de Oliveira, M., Vasconcelos, E., Caminha, C., D'Orleans, J., Belchior, M.: Collective intelligence in law enforcement–The WikiCrimes system. Information Sciences 180, 4–17 (2010)
4. Krötzsch, M., Vrandečić, D., Völkel, M.: Semantic MediaWiki. In: Cruz, I., Decker, S., Allemang, D., Preist, C., Schwabe, D., Mika, P., Uschold, M., Aroyo, L.M. (eds.) ISWC 2006. LNCS, vol. 4273, pp. 935–942. Springer, Heidelberg (2006)
5. Auer, S., Dietzold, S., Riechert, T.: OntoWiki – A Tool for Social, Semantic Collaboration. In: Cruz, I., Decker, S., Allemang, D., Preist, C., Schwabe, D., Mika, P., Uschold, M., Aroyo, L.M. (eds.) ISWC 2006. LNCS, vol. 4273, pp. 736–749. Springer, Heidelberg (2006)
6. Auer, S., Bizer, C., Kobilarov, G., Lehmann, J., Cyganiak, R., Ives, Z.: DBpedia: A Nucleus for a Web of Open Data. In: Aberer, K., Choi, K.-S., Noy, N., Allemang, D., Lee, K.-I., Nixon, L.J.B., Golbeck, J., Mika, P., Maynard, D., Mizoguchi, R., Schreiber, G., Cudré-Mauroux, P. (eds.) ASWC 2007 and ISWC 2007. LNCS, vol. 4825, pp. 722–735. Springer, Heidelberg (2007)
7. Breslin, J.G., Harth, A., Bojars, U., Decker, S.: Towards Semantically-Interlinked Online Communities. In: Gómez-Pérez, A., Euzenat, J. (eds.) ESWC 2005. LNCS, vol. 3532, pp. 500–514. Springer, Heidelberg (2005)
8. McGuinness, D.L., Ding, L., Da Silva, P.P., Chang, C.: Pml 2: A modular explanation interlingua, In: Proceedings of AAAI (2007)

9. Pinyol, I., Sabater-Mir, J., Cuní, G.: How to talk about reputation using a common ontology: From definition to implementation. In: Proceedings of the Ninth Workshop on Trust in Agent Societies, Hawaii, USA. pp. 90–101 (2007)
10. Mollenkopf, J., Goldsmith, V., McGuire, P., Sara, M.L.: Identification, development and implementation of innovative crime mapping techniques and spatial analysis. US Department of Justice 27, Washington, DC (2000)
11. Brickley, D., Miller, L.: FOAF Vocabulary Specification. Technical report, RDFWeb FOAF Project (2003)
12. Pinheiro, V., Pequeno, T., Furtado, V., Franco, W.: InferenceNet.Br: Expression of Inferentialist Semantic Content of the Portuguese Language. In: Pardo, T.A.S., Branco, A., Klautau, A., Vieira, R., de Lima, V.L.S. (eds.) PROPOR 2010. LNCS (LNAI), vol. 6001, pp. 90–99. Springer, Heidelberg (2010)
13. Pinheiro, V., Furtado, V., Pequeno, T., Ferreira, C.: Towards a Common Sense Base in Portuguese for the Linked Open Data Cloud. In: Caseli, H., Villavicencio, A., Teixeira, A., Perdigão, F. (eds.) PROPOR 2012. LNCS, vol. 7243, pp. 128–138. Springer, Heidelberg (2012)
14. Santos, H., Furtado, V., Pinheiro, V., Ferreira, C., Shiki, G., Eurico Filho, J.: Widgets baseados em conhecimento advindo de dados referenciados e abertos na Web. In: Anais WEBMEDIA 2011, Florianópolis, vol. 1. SBC, Campinas (2011)
15. Bizer, C., Cyganiak, R.: D2r server-publishing relational databases on the semantic web. In: 5th International Semantic Web Conference, p. 26 (2006)
16. Corlosquet, S., Delbru, R., Clark, T., Polleres, A., Decker, S.: Produce and Consume Linked Data with Drupal! In: Bernstein, A., Karger, D.R., Heath, T., Feigenbaum, L., Maynard, D., Motta, E., Thirunarayan, K. (eds.) ISWC 2009. LNCS, vol. 5823, pp. 763–778. Springer, Heidelberg (2009)
17. Auer, S., Lehmann, J., Hellmann, S.: LinkedGeoData: Adding a Spatial Dimension to the Web of Data. In: Bernstein, A., Karger, D.R., Heath, T., Feigenbaum, L., Maynard, D., Motta, E., Thirunarayan, K. (eds.) ISWC 2009. LNCS, vol. 5823, pp. 731–746. Springer, Heidelberg (2009)
18. Borsje, J., Levering, L., Frasincar, F.: Hermes: a semantic web-based news decision support system. In: Proceedings of the 2008 ACM Symposium on Applied Computing, pp. 2415–2420 (2008)

User-Centric Principles in Automated Decision Making

Ingrid Nunes[1,2], Simon Miles[2], Michael Luck[2], and Carlos J.P. de Lucena[1]

[1] LES, Departamento de Informática, PUC-Rio, Rio de Janeiro, Brazil
{ionunes,lucena}@inf.puc-rio.br
[2] Department of Informatics, King's College London, London, WC2R 2LS, United Kingdom
{simon.miles,michael.luck}@kcl.ac.uk

Abstract. Natural-language preference expressions, not yet exploited by existing preference reasoning approaches, match the way users express preferences in many scenarios and potentially improve automated decision making. Further, the preferences provided are often not sufficient to make a choice on behalf of users, as trade-offs are resolved with psychological processes employed in light of available options. We thus propose a decision making technique that reasons about preferences expressed in a user-centric language and incorporates principles of trade-off contrast and extremeness aversion, as in human decision-making.

Keywords: Decision making, Preference Reasoning, Human Reasoning.

1 Introduction

Many everyday tasks involve decisions over a large number of options [10]: we must decide which clothes to wear, what to eat, where to go for fun. Both these regular decisions, and irregular ones, e.g. planning vacations, demand an effort that can be reduced by delegating decision-making to intelligent agents. For agents to appropriately perform tasks on our behalf, however, they must be aware of user preferences and the options available. While existing work allows agents to reason about a restricted set of preference types, this constrains users in how they express preferences and requires tedious interactive elicitation methods. We thus propose a novel approach to reasoning about preferences. Specifically, our contributions are: an automated *decision-making technique* based on preferences expressed in a *high-level preference language* and available options; exploiting principles regarding the way in which humans make choices; and an *evaluation* that compares decisions made by our technique with a human expert.

Our goal with the proposed technique is to simulate human reasoning in making decisions, allowing us to exploit natural user expressiveness of preferences (without the need for elicitation methods) and resolve trade-offs (that cannot be resolved with the provided preferences) in a way humans would do if provided with sufficient time and knowledge. We thus apply psychological processes used by humans, including mainly the principles of *trade-off contrast* and *extremeness aversion* [11]. Our decision-maker takes as input a set of options over which a choice is made, and a preferences expressed in a high-level language. It processes these to select one option, in such a way that the choice can be justified from the preferences. A decision here concerns choosing one option from a set *Opt* of the same conceptual class, e.g. apartments. Each class has a

L.N. Barros et al. (Eds.): SBIA 2012, LNAI 7589, pp. 42–51, 2012.

Table 1. Preference Language (EBNF)

$preference ::= [condition] (constraint|goal|order|qualifying|rating|indifference|dontCare)$
$condition ::= \textbf{if } constraint \textbf{ then}$
$constraint ::= expression \mid expression \textbf{ and } expression \mid expression \textbf{ or } expression \mid \textbf{not } expression$
$expression ::= attribute (= \mid \neq \mid > \mid \geq \mid < \mid \leq) value$
$goal ::= (\textbf{minimise} \mid \textbf{maximise}) attribute$
$order ::= attribute = value \succ attribute = value$
$qualifying ::= performative\ constraint$
$rating ::= constraint\ rate$
$indifference ::= \textbf{indifferent } constraint \{constraint\}$
$dontCare ::= \textbf{dont\_care } attribute$
$performative ::= [\textbf{dont}] (\textbf{prefer} \mid \textbf{need} \mid \textbf{desire} \mid \textbf{avoid} \mid \textbf{like} \mid \textbf{want} \mid \textbf{accept} \mid \textbf{require} \mid \textbf{love} \mid \textbf{hate})$
$rate ::= \textbf{best} \mid \textbf{v\_good} \mid \textbf{good} \mid \textbf{neutral} \mid \textbf{bad} \mid \textbf{v\_bad} \mid \textbf{worst}$
$priority ::= [condition] (attribute\_priority \mid attribute\_indifference \mid preference\_priority)$
$attribute\_priority ::= attribute \triangleright attribute$
$attribute\_indifference ::= attribute \sim attribute$
$preference\_priority ::= \mathbb{Z}. preference$

set *Att* of attributes, e.g. price, and each att_i is associated with a domain D_i that: (i) comprises a set of values x_{ij} allowed for the attribute; (ii) can be *discrete* or *continuous*; and (iii) can be *ordered* or *non-ordered*. As humans express preferences in many ways, we propose a preference language (Table 1), which is composed of different types of preferences and priorities and was derived from a previous study [9]. To illustrate, suppose *Bob* is visiting a university, and needs to choose an apartment to stay at. Each apartment is described in terms of four attributes: (i) distance from the university (**uni**); (ii) distance from the nearest underground station (**station**); (iii) chain (**chain**); and (iv) price per week (**price**). Bob's preferences are shown in Table 2a, with a prioritisation of attributes on the final line. The apartments available are in Table 2b.

Our technique initially analyses the options with regard to preferences, building two models for future use. As some preferences include important *implicit* information, in addition to their literal meaning, we extract this also. From the set of available options, we eliminate those that do not meet strict constraints, or are dominated in every regard by other options (though this latter step is not detailed in this paper due to space restrictions). As the remaining options have both costs and benefits, we take into account relative importance of attributes, and then go beyond the provided preferences with the user-centric principles, concluding with a decision. We make a few limiting assumptions: preferences are consistent (but may conflict); decisions do not concern *critical* matters, where a wrong choice may have serious consequences; decisions concern choice from a finite set of options; and, each preference (excluding its conditions) refers to only one attribute. The technique may be seen as a framework as it has variable parts, which were instantiated in this paper after running the technique with different alternatives, but our future work is to improve results by exploring this variability.

2 Models to Support Decision Making

2.1 Preference Satisfaction Model

Performatives such as *need* and *like* are widely adopted by users to express preferences over attributes, and so are included in our language. Similarly, users may rate

Table 2. Running example: preferences, options and decision models

(a) Preferences.

1. **prefer** $uni \leq 2.5Km$
2. **if** $uni \leq 2.5Km$ **then need** $station \leq 1Km$
3. **if** $uni > 2.5Km$ **then need** $station \leq 0.7Km$
4. **minimise** $station$
5. **minimise** $price$
6. **avoid** $chain = D$
7. $chain = A \succ chain = B$
8. $chain = B \succ chain = C$
if $uni > 2.5Km$ **then** $station \rhd uni$

(b) Set of available apartments.

Attribute	Domain	Ap_A	Ap_B	Ap_E
Uni (Km)	$\{x \mid x \in \mathbb{R}, x > 0\}$	5.5	2.0	3.5
Station (Km)	$\{x \mid x \in \mathbb{R}, x > 0\}$	0.40	0.45	0.65
Chain	$\{A, B, C, D\}$	C	D	A
Price ($)	$\{x \mid x \in \mathbb{R}, x > 0\}$	100	130	100

(c) PSM.

	uni	station	chain	price
Ap_A	¬ prefer	need	¬ avoid	
Ap_B	prefer	need	avoid	
Ap_E	¬ prefer	need	¬ avoid	

(d) OAPM.

	Ap_A		Ap_B		Ap_E	
	Ap_B	Ap_E	Ap_A	Ap_E	Ap_A	Ap_B
uni	-	~	+	+	~	-
station	+	+	-	+	-	-
chain	+	-	-	-	+	+
price	+	~	-	-	~	+

preferences from best to worst. The relative importance of performatives, and their relation to rates, is specific to each individual, and eliciting this is outside the scope of this paper. Here, we adopt one ranking and categorisation as positive, negative and neutral, shown in Figure 1. Rates and performatives (collectively referred to as *modifiers*, M) used in preferences are captured by a *Preference Satisfaction Model (PSM)*, a table indicating how options satisfy preferences in terms of each attribute. This maps a pair $\langle option, attribute \rangle$ to a performative or rate (or their negation): $PSM : Opt \times Att \rightarrow \{empty, \neg\} \times M$. Modifiers come from qualifying and rating preferences but also constraints, interpreted as the performative "*want.*". Thus, each preference considered in the PSM consists of a modifier, a constraint, and, optionally, a condition.

The PSM is constructed as follows. For each preference, the relevant attribute of each option is tested to see if the preference is satisfied (condition and constraint met). For each option-attribute pair, the modifier from one preference is chosen as the respective value in the PSM as follows. If at least one constraint is satisfied, the strongest modifier according to our scale is chosen, either the most positive or most negative (depending on whether the modifier is a positive or negative one). If no statement is satisfied, the least negative modifier is chosen, and marked negated in the PSM. Table 2c shows the PSM for our running example.

2.2   Options-Attribute Preference Model

To capture comparisons between options, we introduce the Options-Attribute Preference Model (OAPM), a table that, for individual attributes, shows which of each pair

+	require > need > love/want/best > desire/v_good > prefer/like/good > accept
+-	don't require > neutral/don't love/don't hate > dont't need/don't avoid/don't desire
-	don't prefer > avoid/bad > don't like/v_bad > don't want/worst > hate > don't accept

Fig. 1. Modifier strength scale (performatives and rates)

of options is better. Four possible values relate each option A to an option B w.r.t. an attribute X: (i) ($+$) the attribute value of A is better than B; (ii) ($-$) A is worse than B; (iii) ($\sim$) A is as preferred as B; and (iv) (?) no conclusion can be drawn with the provided preferences. This is represented as $OAPM : Opt \times Opt \times Att \rightarrow \{+, -, \sim, ?\}$. The result for our running example is shown in Table 2d. The OAPM is initialised with all values set to "?". We then analyse the following in order: (i) the PSM; (ii) goals; (iii) order and indifference preferences. The ordering allows each step to override or refine values derived from prior steps.

PSM. We use our modifier scale (Figure 1) to determine the preferred value of two options' attributes, following the rules specified below, shown as situations in which the PSM establishes preference of o_1 over o_2. Note that an absence of satisfied preferences for an option, $\langle \neg, modifier \rangle$, is considered an indicator that the option is undesirable, following typical practice that people explicitly state where attribute values are desired, acceptable, not to be avoided etc. [10]. In our running example, $OAPM(Ap\_A, Ap\_B, uni)$ is set to $-$, as $PSM(Ap\_A, uni) = \neg prefer$ and $PSM(Ap\_B, uni) = prefer$.

o_1	$\langle empty, stronger + m \rangle$	$\langle empty, + m \rangle$	$\langle empty, \pm m \rangle$	$\langle empty, \pm m \rangle$	$\langle \neg, m \rangle$
o_2	$\langle empty, m \rangle$	$\langle \neg, m \rangle$	$\langle \neg, + m \rangle$	$\langle \neg, \pm m \rangle$	$\langle empty, - m \rangle$

Goals. When there is a goal to maximise the attribute, the option with a higher value is set to $+$, and the other set to $-$. The inverse applies for a minimisation goal. According to the PSM, Ap_A and Ap_E are similar w.r.t. *station* (both are associated with $\langle empty, need \rangle$), but as there is a goal to minimise this attribute, the OAPM is updated to $OAPM(Ap\_A, Ap\_E, station) = +$ and $OAPM(Ap\_E, Ap\_A, station) = -$.

Order Preferences. Order preferences state a partial order between values of an attribute, where different orders may apply under different conditions. We create a graph, an *attribute value partial order (AVPO)*, for each option-attribute pair, whose nodes contain equality expressions of order preferences (which represent domain values), while directed edges denote the preference of one value to another. For example, preferences 7 and 8 of our running example results in the following AVPO: $chain = A \rightarrow chain = B \rightarrow chain = C$ — for all options, as there are no conditions. For a given attribute and options o_1 and o_2, where there is a path from the node that satisfies o_1's value to that satisfying o_2's value in both options' AVPO, then this means the orders applicable to each option both consider o_1 preferable to o_2 for that attribute, and the OAPM entries are set to $+$ and $-$ accordingly. This is the case with options Ap_E and Ap_A, whose chains are A and C, respectively.

Indifference Preferences. The OAPM value of an option-attribute pair is set to $\sim$ if the attribute values satisfy at least one constraint of the same indifference preference, and its condition (if any) is satisfied by both options.

Preferences always provide a literal meaning, but can also bring additional information to derive new preferences, referred to as *implicit preferences*. These never override explicitly provided preferences, but aid ordering attribute values when this is otherwise inconclusive, i.e. where $OAPM(o_1, o_2, att) = \sim \vee$? and this is not due to an explicit indifference preference. We consider four kinds of implicit preference in refining the

OAPM. First, when an **upper bound** is specified for an attribute, we assume that this implies a goal to minimise the attribute value. For example, a user expecting to pay at most $100 for a hotel night also wants to minimise price. Conversely, a **lower bound** implies a goal to maximise the attribute value. A **reference value (around preference)** implies a goal that a value closer to the reference is better. Finally, an **interval** of acceptable values ($att > lowerLimit \land att < upperLimit$) implies, for values outside that interval, a goal of being closer to the nearest interval boundary. For all such implicit preferences, if the modifier associated with the qualifying or rating preference is negative, the effect is inverted. In our example, preference 1 suggests an implicit preference for *minimising* the value of the *uni* attribute.

3   Cost and Benefit Analysis

We next assess the relative costs and benefits of pairs of options. We first analyse the benefits of option o_1 compared to option o_2 for each attribute, and the same for o_2 compared to o_1. The benefits of option o_1 w.r.t. o_2 are the costs of option o_2 w.r.t. o_1. $Benefit : Opt \times Opt \times Att \to \{x | x \in \mathbb{R} \land 0 \geq x \geq 1\}$ indicates how much better one option is than another, w.r.t. to one attribute. If the $OAPM(o_1, o_2, att) \neq +$, then the benefit is 0, otherwise, $Benefit(o_1, o_2, att)$ is computed based on the *reason*, preference or PSM value, for setting the final OAPM value as described above. Our decision maker keeps track of these reasons when building the OAPM. Different cases are considered depending on the reason.

First, the reason may be a goal or an implied preference of a kind described above. If a **goal, upper or lower bound**, the benefit is the difference between the options' attribute values. If an **around preference**, the benefit is the difference between the distances of options' values from the reference value. If an **interval preference**, the benefit is the difference between the distances of options' values from the nearest interval boundary. In all these cases, the difference is normalised to $[0, 1]$ within the minimum and maximum domain value bounds of the attribute (which are given, or we extract from the the available options).

If the reason is **PSM values**, we assess how much one value is preferred to another. We order our modifiers, and associate a numeric value with each level of the scale, with 0 for the middle level (*neutral, don't love, don't hate*), increasing 1 for each level above and reducing by 1 for each level below. The absolute benefit for PSM value $\langle n, m \rangle$ with modifier m at level *level* is as follows: $f_m(level) = \log(|level| + 1)$, if $level \geq 0$, $n = empty$; $f_m(level) = -\log(|level| + 1)$, if $level < 0$, $n = empty$; and $f_m(level) = 0, n = \neg$. The relative benefit between options is then the difference between the absolute benefits of the options. We use a logarithmic function above so that differences between stronger modifiers, such as *require*, are less than differences between modifiers in the middle of the scale, such as *don't avoid*, consequently the preference is much stronger when comparing positive modifiers with negative modifiers.

If the reason is an **order preference**, then the AVPO graphs are used to calculate the benefit. Each AVPO node, corresponding to an attribute value, is tagged with a modifier according to the same algorithm used to construct the PSM for the options, i.e. whether that value is preferred, not avoided etc. Nodes with only incoming or outgoing edges that cannot be tagged are given default modifiers, *want* and *prefer* respectively,

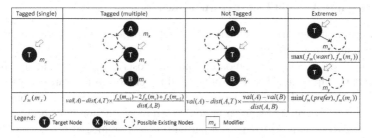

Fig. 2. Calculating order node values

or stronger modifiers if there are other tagged nodes with more positive or negative modifiers than the default. Less preferred nodes are tagged with *prefer* by default, since people typically provide an order for preferred or acceptable values, and ignore others. Each node is then given absolute benefit value as follows (some details omitted due to space restrictions). If a modifier is tagged to only one node, then its benefit is calculated using f_m above. If multiple nodes have the same modifier, with level *level*, their benefits are an even distribution from $f_m(level) - (f_m(level) - f_m(level - 1)/2)$ to $f_m(level) + (f_m(level + 1) - f_m(level)/2)$ following the edge order in the AVPO. Untagged nodes' benefits are determined from those of connected nodes. Figure 2 illustrates the calculation. The relative benefits between options are then calculated as the difference between absolute benefits of the nodes containing the two options' values in each AVPO, normalised to [0,1] by the minimum and maximum values of the modifier scale. The values from the two AVPOs are then averaged to reach the final benefit value. Nodes of the *chain* AVPO of our example are tagged with $\neg avoid$, and their values are calculated with the equation of multiple tagged nodes.

Benefits of options across all attributes are calculated by considering the prioritisation of preferences and attributes expressed with the priorities. First, for each option, the preference priority order (1 to 8 in our example) gives an initial ordering of attributes: attributes constrained by higher priority preferences whose conditions hold for that option are more important, e.g. *Ap_A* gives order *uni* ≻ *station* ≻ *price* ≻ *chain*. Next, we consider those explicit attribute prioritisations and indifference whose conditions hold, and change the order accordingly, e.g. for *Ap_A*, *station* and *uni* are swapped. Finally, don't care preferences indicate attributes to be removed from the order. Given this attribute order, we take the least important attributes to be level 1 in the order, and the longest path in the order from the least important attributes to the most important ones is referred to as $size(attO)$. We use a logarithmic function ($f_a(x) = a \log x + b$) to calculate attribute weights when considering the overall option benefits, with $f_a(1) = 1$ and $f_a(size(attO)) = size(attO)$. We then calculate parameters a and b, according to the number of levels. The logarithmic function, with characteristics imposed by the points established, gives a much higher priority to more important attributes, which have similar importance (in comparison to a linear function). Based on the logarithmic function with specific parameters, $w_i = f_a(level(i))/\sum_{j=1}^{size(att)} f_a(level(j))$ calculates the weight of each attribute $w_i \in Att$. Now that we have the benefits of an option o_1 w.r.t. an option o_2, for each individual attribute, and also its weights, we calculate the overall benefits from o_1 w.r.t. o_2 using a weighted sum.

3.1 Taking into Account User-Centric Principles

As we are not considering dominated options in this paper, options have both pros and cons, and thus a trade-off must be resolved to choose one of them. People not only consider the two options being compared, and their costs and benefits, but also the cost-benefit relationship (ratio), which is positioned in relation to this ratio between other options [11]. This is referred to as *trade-off contrast*. We therefore incorporate a new factor in the process of choosing an option, based on a function that shows the trade-off between two options TO. This is a partial function defined as $b(o_2, o_1)/b(o_1, o_2)$, whose domain is every pair of different options for which $b(o_1, o_2) > b(o_2, o_1)$. As the function b is always a value in the interval $[0, 1]$, the TO function is always a value in this interval, excluding its boundaries. The trade-off between two options is not isolated; with only two options, all we know is that one option has more benefits than another. When there are other options, and the (human) decision-maker observes that the cost-benefit relationship is better for other options, they see it as a negative aspect of the option. This counters *rational* decision-making, as preferences for a particular option do not depend on available options. Based on the TO function, the option benefits w.r.t. trade-off are as below, having as a basis the avg_{TO} (average of all values defined for the TO), which determines when the trade-off is a benefit or a cost.

$$to(o_1, o_2) = \begin{cases} avg_{TO} - TO(o_1, o_2) & \text{if } TO(o_1, o_2) \text{ is defined and } TO(o_1, o_2) \leq avg_{TO} \\ TO(o_2, o_1) - avg_{TO} & \text{if } TO(o_2, o_1) \text{ is defined and } TO(o_2, o_1) > avg_{TO} \\ 0 & \text{otherwise} \end{cases}$$

Humans also consider how *extreme* options are. Extreme options are close to best for some attributes, e.g. quality, but incur a high penalty for others, e.g. price. In general, humans avoid extreme options [11], referred to as *extremeness aversion*. To evaluate how extreme options are, we calculate the distance between an option's attribute values from the best possible value according to preferences whose conditions are valid for the option ($bestDist(o, att)$). The precedence order for using preferences to calculate the distance from best is: (i) goals and implicit preferences, (ii) order, and (iii) PSM. An attribute is considered only if there is no *don't care* preference associated with it. Extreme options have low costs for some attributes ($bestDist$ close to 0) and high costs for others ($bestDist$ close to 1), so extremeness of an option is given by the standard deviation of the function $bestDist$ for a particular option: $ext(o) = STDEV(\{bestDist(o, att_i)|i = 1...size(att)\})$, which is a value between 0 and 1. Finally, extremeness aversion suggests that a less extreme option has a benefit with respect to a more extreme option. To capture this aspect, we define the function $ea(o_1, o_2)$, which is $ext(o_2) - ext(o_1)$ if $ext(o_1) < ext(o_2)$, and 0 otherwise.

3.2 Comparing Option Relative Values

We have analysed three aspects of options: benefits, trade-off relative to available options, and extremeness. The last two aspects are also seen as benefits: trade-offs better than the average are also a benefit, and the the least extreme of two options has a benefit w.r.t. the other. The final value $v(o_1, o_2)$ of an option is a weighted sum of these aspects — we are currently using default weights of 0.25 for trade-off contrast and 0.15 for extremeness aversion. We identify the chosen option as better than or equal

Table 3. Running Example: choosing an option

	uni	$station$	$chain$	$price$	$b(o_1,o_2)$	$ea(o_1,o_2)$	$TO(o_1,o_2)$	$to(o_1,o_2)$	$v(o_1,o_2)$	**Balance**
Ap_A	0.0	0.031	0.031	0.230	0.291	0.227	0.421	0.320	0.289	**0.216**
Ap_B	0.123	0.0	0.0	0.0	0.123	0.0		0.0	0.073	
Ap_B	0.123	0.103	0.0	0.0	0.226	0.0		0.126	0.167	
Ap_E	0.0	0.0	0.031	0.230	0.260	0.144	0.868	0.0	0.178	**0.011**
Ap_A	0.0	0.153	0.0	0.0	0.153	0.084		0.194	0.153	**0.055**
Ap_E	0.124	0.0	0.040	0.0	0.164	0.0	0.935	0.0	0.098	

to $(v(o_1,o_2) \geq v(o_2,o_1))$ every other option. If different options have the same value w.r.t. another $(v(o_1,o_2) = v(o_2,o_1))$, and they are better than every other option, we randomly choose one of them. Cycles may arise in the v function if different criteria are used to compare attribute values of each option pair; e.g., the price of o_1 and o_2 is compared by goal, and of o_1 and o_3 by constraint. As preferences are consistent, price alone gives no cycles; however, as v is calculated for different preferences, differences in scales can lead to cycles when considering overall option benefits. To choose one option in this situation, from the set of options that are considered better than the highest number of options, we choose that with the minimum of the maximum balances for every option that is considered better than it. In our experiments, on real user data, there were only 16 (of 113) occurrences of cycles. Even though this number is low and we use a workaround to solve this issue, it is future work to completely eliminate cycles.

In our example, our technique results in the values presented in Table 3. Here, the **chosen option is Ap_A**, which would not have been chosen without our user-centric principles. As already introduced, our technique goes beyond provided preferences, because they do not give enough information to resolve trade-offs, which humans do *during* decision-making. Our technique aims to anticipate this preference construction in order to make a decision on behalf of the user or provide a recommendation.

4   User Study-Based Evaluation

We evaluate our decision-maker by empirical evaluation but, as the input of our technique is high-level preferences and existing approaches cannot handle all of them, we restrict ourselves to making a side-by-side comparison with a human expert. The evaluation is based on the study [9] that also informed the preference language itself. Participants provided preference specifications (in natural language) for use by an individual to buy a laptop on their behalf. Both the participants and domain expert (based on the participant's preferences), were given a catalogue with 144 laptops from which to choose up to five options. We compared decisions made by our technique based on provided preferences against those of the user and expert. Similarly to how the expert recommendation was assessed in the user study, we calculate a similarity score SS (which ranges from 0 to 100), comparing the recommendation with the user choice and taking into account the position of the up to five chosen laptops using a weighted average.

We run our technique with 113 preference specifications — keeping only keywords to be consistent with our language — as input (taking an average of 10.2832 *seconds* on an Intel Core 2 Quad $2.66GHz$, $4GB$ of RAM, with standard deviation 0.6465,

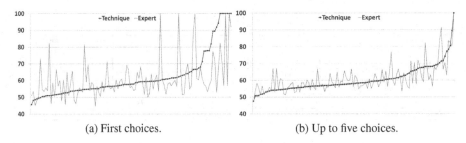

(a) First choices.   (b) Up to five choices.

Fig. 3. Expert vs technique

Table 4. Reasoning Approaches vs. Preferences

Approach	Preference								Attribute Priority	Attribute Indiff.	Preference Priority
	Condition	Const.	Goal	Order	Qualif.	Rating	Indiff.	Don't care			
UF-based [8]				X							
SVM-based [6]				X		X	X				
Soft Constraints [2]		X									X
CP-Nets [3]	X			X							
TCP-Nets [4]	X			X					X		
Scoring Function [1]		X				X		X			X
Winnow [5]		X		X			X				X

to be executed for each request, with 144 laptops, and 61 attributes), and obtained the similarity scores shown in Figure 3a (first expert and technique choices compared to the first user choice) and 3b (first up to five expert and technique choices compared to the first up to five user choices). Specifically, in the first choice comparison, the average SS of our technique is 61.94 ($stdev = 13.24$, $min = 45.69$, $max = 100.0$), while of the expert is 60.79 ($stdev = 12.05$, $min = 44.72$, $max = 100.0$). And in the up to five choices comparison, the average SS of our technique is 60.36 ($stdev = 7.69$, $min = 47.56$, $max = 100.0$), while of the expert is 61.03 ($stdev = 8.38$, $min = 46.97$, $max = 96.39$). The small difference in obtained values is not statistically significant, as determined by a paired-samples t-test: $t(0.916) = 0.361, p = 0.05$ (F) and $t(-1.268) = 0.207, p = 0.05$ (5). Therefore, we cannot reject the null hypothesis that domain expert and technique choices are different, and *we can conclude that our technique makes choices as good as those of the domain expert.*

5   Related Work and Conclusions

Most existing work related to decision making is founded on Multi-Attribute Utility Theory [7], which emphasises the use of multi-attribute preference models based on *utility functions* (UFs). Many approaches [8,6] propose specific models to represent preferences for deriving *utility functions*. Some approaches [2] extend Constraint Satisfaction Problems (CSPs) to incorporate *soft constraints* (that can remain unsatisfied), associating preference with each constraint, and creating an optimisation problem of maximising preference. UFs and CSPs are classical approaches for dealing with preferences and making decisions, but the former are hard to elicit, and the latter deal with over-constrained problems rather than choosing from feasible solutions. A third group of approaches [3,4] proposes new *graphical structures* to represent and reason about preferences. Finally, work in the area of databases proposes *extensions of query languages* [1,5] to incorporate preferences and algorithms to provide query results according to specified preferences. Even though these approaches propose different solutions, they share the common goal of making a preference-based choice. However, they address limited kinds of preferences (Table 4), restricting human natural expression, and cannot make a decision when the preferences themselves do not lead to a single option to be chosen. As this paper is not concerned with preference elicitation methods, they have not been considered.

In this paper, we provided an approach to reasoning about preferences and making decisions. Our technique provides the novelty of exploiting different natural language expressions and user-centric principles in automated decision making, and these can be used as general lessons in this research area: performatives and other expressions give valuable information that can be used to generate low-level preference representations (such as utility functions), and these (and possibly others) user-centric principles can be used to reduce the amount of preferences obtained from users, as they can predict how users would resolve trade-offs. Moreover, these principles of human decision making explain situations in which a decision made by a human is "irrational" according to classical decision theory, and by taking these principles into account, automated systems can make decisions that are more acceptable to users. Short-term future work is to explore variable parts of our technique, such as modifier interpretation, modifier and weight functions, and weights used for trade-off contrast and extremeness aversion.

References

1. Agrawal, R., Wimmers, E.L.: A framework for expressing and combining preferences. In: Proceedings of the 2000 ACM COMAD, pp. 297–306. ACM (2000)
2. Bistarelli, S., Montanari, U., Rossi, F.: Semiring-based constraint satisfaction and optimization. J. ACM 44, 201–236 (1997)
3. Boutilier, C., Brafman, R.I., Domshlak, C., Hoos, H.H., Poole, D.: Cp-nets: a tool for representing and reasoning with conditional ceteris paribus preference statements. J. Artif. Int. Res. 21(1), 135–191 (2004)
4. Brafman, R.I., Domshlak, C., Shimony, S.E.: On graphical modeling of preference and importance. J. Artif. Int. Res. 25, 389–424 (2006)
5. Chomicki, J.: Preference formulas in relational queries. ACM Trans. Database Syst. 28, 427–466 (2003)
6. Domshlak, C., Joachims, T.: Efficient and non-parametric reasoning over user preferences. User Modeling and User-Adapted Interaction 17(1-2), 41–69 (2007)
7. Keeney, R.L., Raiffa, H.: Decisions with Multiple Objectives: Preferences and Value Tradeoffs. Wiley series in probability and mathematical statistics. Wiley (1976)
8. McGeachie, M., Doyle, J.: Utility functions for ceteris paribus preferences. Computational Intelligence 20(2), 158–217 (2004)
9. Nunes, I., Barbosa, S., Lucena, C.: Understanding how users express preferences: a user study. Tech. Rep. CS-2010-19, University of Waterloo, Canada (2010)
10. Schwartz, B.: The paradox of choice: Why more is less. Harper Perennial (2005)
11. Simonson, I., Tversky, A.: Choice in context: Tradeoff contrast and extremeness aversion. Journal of Marketing Research 29(3), 281–295 (1992)

Wearable Computing: Accelerometers' Data Classification of Body Postures and Movements

Wallace Ugulino[1], Débora Cardador[1], Katia Vega[1],
Eduardo Velloso[2], Ruy Milidiú[1], and Hugo Fuks[1]

[1] Pontifical Catholic University of Rio de Janeiro, Informatics Department
[2] School of Computing and Communications, Lancaster University, UK
{wugulino,dcosta,kvega,milidiu,hugo}@inf.puc-rio.br,
e.velloso@lancaster.ac.uk

Abstract. During the last 5 years, research on Human Activity Recognition (HAR) has reported on systems showing good overall recognition performance. As a consequence, HAR has been considered as a potential technology for e-health systems. Here, we propose a machine learning based HAR classifier. We also provide a full experimental description that contains the HAR wearable devices setup and a public domain dataset comprising 165,633 samples. We consider 5 activity classes, gathered from 4 subjects wearing accelerometers mounted on their waist, left thigh, right arm, and right ankle. As basic input features to our classifier we use 12 attributes derived from a time window of 150ms. Finally, the classifier uses a committee AdaBoost that combines ten Decision Trees. The observed classifier accuracy is 99.4%.

Keywords: Human Activity Recognition, Wearable Computing, Machine Learning, Accelerometer.

1 Introduction

With the rise of life expectancy and ageing of population, the development of new technologies that may enable a more independent and safer life to the elderly and the chronically ill has become a challenge [1]. Ambient Assisted Living (AAL) is one possibility to increase independence and reduce treatment costs, but it is still imperative to generate further knowledge in order to develop ubiquitous computing applications that provide support to home care and enable collaboration among physicians, families and patients.

Human Activity Recognition (HAR) is an active research area, results of which have the potential to benefit the development of assistive technologies in order to support care of the elderly, the chronically ill and people with special needs. Activity recognition can be used to provide information about patients' routines to support the development of e-health systems, like AAL. Two approaches are commonly used for HAR: image processing and use of wearable sensors.

The image processing approach does not require the use of equipment in the user's body, but imposes some limitations such as restricting operation to the indoor

L.N. Barros et al. (Eds.): SBIA 2012, LNAI 7589, pp. 52–61, 2012.
© Springer-Verlag Berlin Heidelberg 2012

environments, requiring camera installation in all the rooms, lighting and image quality concerns and, mainly, users' privacy [2]. The use of wearable sensors minimizes these problems, but requires the user to wear the equipment through extended periods of time. Hence, the use of wearable sensors may lead to inconveniences with battery charges, positioning, and calibration of sensors [3].

We performed a literature review on HAR using wearable accelerometers data, discussed in Section 2. This review unfolded an intense growth on the number of publications related to wearable accelerometers in the last 5 years. It was further observed that few works provide public datasets for benchmarking and there is no sufficient information on the specification and orientation of the sensors deployed. The absence of public datasets and information for the reproduction of the studies are evidences of the need for maturity within the area. In view of such results, it was decided for the investigation of activities recognition by means of wearable accelerometers approach.

In this project we built a wearable device with the use of 4 accelerometers positioned in the waist, thigh, ankle and arm. The design of the wearable, details on the sensors used, and other necessary information for the reproduction of the device are shown on Section 3. We collected data from 4 people in in different static postures; and dynamic movements with which we trained a classifier using the AdaBoost method and decision trees C4.5 [3, 5]. The design of the wearable, data collection, extraction and selection of features and the results obtained with our classifier are described in Section 4.   Conclusion and future work are discussed in Section 5.

2   Literature Review

The results presented in this section are part of a more comprehensive systematic review about HAR with wearable accelerometers. The procedures used for the results of this paper are the same used in a traditional systematic review: we defined a specific research question, used a search string in the database, applied exclusion criteria and reviewed resulting publications in qualitative and quantitative form. For the quantitative analysis, we collected metadata from articles and used descriptive statistics to summarize data. The method application is described as follows:

- Research Question: What are the research projects conducted in recognition of human activities and body postures with the use of accelerometers?
- Search string: (((("Body Posture") OR "Activity Recognition")) AND (accelerometer OR acceleration)). Refined by: publication year: 2006 – 2012;
- Results in IEEE database: 144 articles;
- Exclusion criteria:
    - Use of accelerometers in smartphones;
    - HAR by image processing;
    - Not related to human activity (robots, in general);

- Different task (composite activities, games, gesture input recognition, etc.);
- Researches in the area of sensors network, without development of predictors/classifiers;
- It was used the most recent publication when the same result was published in different moments;
- Research about energy consumption in wearable devices;
• Result: 69 articles for quantitative and qualitative analysis.

For the quantitative analysis the metadata drawn from the articles were as follows: research title, year, quantity of accelerometers, use of other sensors, accelerometers position, classes, machine learning technique (or threshold based algorithms), number of subjects and samples, test mode (training dataset + test dataset, cross-validation with modes: k-fold, leave-one-example-out, or leave-one-subject-out), percentage of correctly classified samples. It was observed in relation to the publication year, that there is a growing number of publications on HAR with Wearable Accelerometers, as shown in Figure 1, which shows evidence of the importance of the approach for the Human Activity Recognition community.

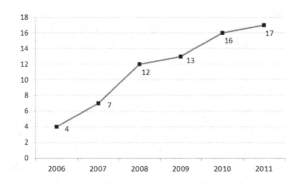

Fig. 1. IEEE publications based on wearable accelerometers' data for HAR

In the surveyed works, we observed the use of up to 4 accelerometers in the collection of data for the most part of the works. The most widely used test mode is the k-fold cross-validation; however, less-dependable tests and even non-standard tests were performed in some recent works. Another important data identified in the literature is that most works present a percentage starting at 90% of success rate in the activities' classification. However, in just 7 works it was informed the dataset size. A list of metadata drawn from the most recent publications (2012 and 2011) is shown on Table 1.

Table 1. HAR based on accelerometers' data from 2012 and 2011 (IEEE database)

Research	# of sensors	Accelerometers' position	Solution	# of users	Learning mode	Test mode	Correct (%)
Liu et. al. (2012) [6]	1	hip, wrist (no info about orientation)	SVM	50	Super-vised	leave-one-out	88.1
Yuting et al. (2011) [7]	3	chest and both thighs (no info about orientation)	Threshold-based	10	--	--	98.6
Sazonov et al. (2011) [8]	1	foot	SVM	9	Super-vised	4-fold cross validation	98.1
Reiss & Stricker (2011) [9]	3	lower arm, chest and foot	Boosted Deci-sion Tree	8	Super-vised	8-fold cross validation	90.7
Min et al., (2011) [10]	9	torso, arms and legs	Threshold-based	3	--	Comparison with k-means	96.6
Maekawa & Watanabe (2011) [11]	4	wrists of both hands, waist, and right thigh	HMM	40	Unsu-pervised	leave-one-out	98.4
Martin et al. (2011) [12]	2	hip, foot and chest	Threshold-based	5	--	--	89.4
Lei et al. (2011) [3]	4	waist, chest, thigh, and side of the body	Naïve Bayes	8	Super-vised	Several, w/ no cross validation	97.7
Alvarez et al. (2011) [13]	1	centered in the back of the person	Genetic fuzzy finite state machine	1	Super-vised	leave-one-out	98.9
Jun-ki & Sung-Bae (2011) [14]	5	forehead, both arms, and both wrists	Naïve Bayes and SVM	3	Super-vised	leave-one-out	**99.4**
Ioana-Iuliana & Rodica-Elena (2011) [15]	2	right part of the hip, lower part of the right leg	Neural Networks	4	Super-vised	66% trainning vs 33% test	**99.6**
Gjoreski et al. (2011) [2]	4	chest, waist, ankle and thigh	Naïve Bayes, SVM, C4.5, Random Forest	11	Super-vised	Leave-one-person-out	90.0
Feng, Meiling, and Nan (2011) [16]	1	Waist	Threshold-based	20	--	--	94.1
Czabke, Marsch, and Lueth (2011) [17]	1	Trousers' Pocket	Threshold-based	10	--	--	90.0
Chernbumroong, et al. (2011) [18]	1	Non-dominant wrist (watch)	C4.5 and Neural Networks	7	Super-vised	5-fold cross-vali-dation	94.1
Bayati & Chavarriaga (2011) [19]	--	Simulations instead of real accelerometers	Expectation Maximization	--	Unsuper-vised	Not mentioned	86.9
Atallah et al (2011) [20]	7	ear, chest, arm, wrist, waist, knee, and ankle	Feature Selection algorithms*	11	Super-vised	Not applied	--
Andreu et al. (2011) [21]	1	Not mentioned	fuzzy rule-based	--	Online learning	--	71.4

* This work is about sensor positioning and feature extraction

In the set of articles assessed, it was observed that the subject independent analysis has been less explored: just 3 out of 69 articles presented a subject independent analysis. The primary alternatives presented by the authors in order to improve the prediction performance in subject independent tests are: (1) increase of dataset, performing the data collection from subjects of different profiles; (2) adapting learning to a subject from data collected from subjects with similar physical characteristics [11]; and (3) investigation of subject independent features and more informative of the classes [22].

Among the articles assessed, we observed a discussion on the importance of the location of the accelerometers on the body. The positions in which wearable accelerometers most commonly mounted on are the waist, next to the center of mass, and chest. A research investigates specifically the development of classifiers adaptable to different amounts of accelerometers [20].

It was also observed that the main problematic issue found was the unavailability of the dataset, which restrains the comparison of results between researches. There is also a lack of information about the orientation of the sensors axis in most of the researches, although the location is usually well described. In some research it was not informed the model of sensor used. The absence of information about orientation and sensors model impairs the reproduction of the wearable devices.

As the area matures, it is imperative that datasets be published in order to enable comparison of results. The importance of disclosing datasets for benchmarking is also commented by Yang & Lianwen [23] as it acknowledges that "*the recognition algorithms rely heavily on the dataset*". The authors also verified the unavailability of datasets in the area and denounced the existence of unpublished proprietary datasets. As a form of contribution, the authors made available a dataset for benchmark in HAR, in addition to details on the location, orientation and model of sensors used. The dataset provided by the authors is small, with only 1278 samples, but it is an important step towards the maturity of this research area. It is necessary that more datasets are publicized, about different activities, and that accurate information on the model, positioning and orientation of sensors are provided in order to enable the comparison of research results in the HAR area by means of wearable accelerometers.

In order to enable the reproduction of the literature review discussed in this article, all publications assessed in this paper are available in RIS format, in the following web address: http://groupware.les.inf.puc-rio.br/har. In this research, the bibliographic management and publishing solution used was the EndNote X5™. The library in EndNote format is also available in this project web address.

3   Building Wearable Accelerometers for Activity Recognition

Our wearable device comprised 4 tri-axial ADXL335 accelerometers connected to an ATmega328V microcontroller. All modules were of the Lilypad Arduino toolkit. The wearable device and the accelerometers' positioning and orientation diagram are illustrated in Figure 2.

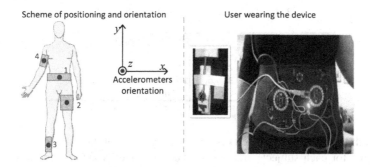

Fig. 2. Wearable device built for data collection

The accelerometers were respectively positioned in the waist (1), left thigh (2), right ankle (3), and right arm (4). All accelerometers were calibrated prior to the data collection. The calibration consists of positioning the sensors and the performance of the reading of values to be considered as "zero". From the calibration, the read values of each axis during data collection are subtracted from the values obtained at the time of the calibration.

The purpose of the calibration was to attenuate the peculiar inaccuracy issues of this type of sensor. Because of this, the sensors were calibrated on top of a flat table in the same position. Another regular type of calibration is the calibration by subject [3], in which the accelerometers are read and calibrated after positioned in the subjects' bodies. The calibration by subject may benefit the data collection provided that it enables the obtainment of more homogeneous data. However, it makes the use of the wearable after completion more complex.

Even though the Lilypad Arduino platform was design to be used with conductive thread, we used wired cablesin order to increase robustness, to enable on-the-fly changes in the circuit and to facilitate reprogramming the microcontroller. The assembly took into consideration the ease of move by the users. The source code (sketch in the *Arduino* Programming Language) is available at http://groupware.les.inf.puc-rio.br/har.

4   Building a Classifier for Wearable Accelerometers' Data

We took the following steps to develop a classifier for the data achieved from the 4 accelerometers: data collection, data pre-processing, feature extraction, feature selection, and 10-fold cross-validation type tests to assess the accuracy of the classifier developed.

4.1   Data Collection

We collected data during 8 hours of activities, 2 hours with each one of the 4 subjects: 2 men and 2 women, all adults and healthy. The protocol was to perform each activity separately. The profile of each subject is shown in Table 2:

Table 2. Characteristics of the participants

Subject	Genre	Age	Height	Weight	Instances
A	Female	46 y.o.	1.62m	67kg	51,577
B	Female	28 y.o.	1.58m	53kg	49,797
C	Male	31 y.o.	1.71m	83kg	51,098
D	Male	75 y.o.	1.67m	67kg	13,161*

* A smaller number of observed instances because of the participant's age

Although the number of subjects is small, the amount of data collected is reasonable (2 hours for each subject) and the profile is diverse: women, men, young adults and one Elder. At total it was collected 165,633 samples for the study; the distribution of the samples between the classes is illustrated in Figure 3.

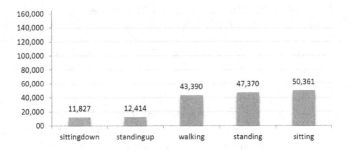

Fig. 3. Frequency of classes between collected data

4.2   Feature Extraction

From the data collected from the tri-axial accelerometers it was performed a data pre-processing, following some instructions from [24]. It was generated a 1 second time window, with 150ms overlapping. The samples were grouped and descriptive statistic was used for generating part of the derivate features. The derivate features of acceleration in axis x, y, and z and of the samples grouped are listed as follows:

- For each accelerometer: Euler angles of roll and pitch and the length (module) of the acceleration vector (called as total_accel_sensor_n);
- Variance of roll, pitch and module of acceleration for all samples in the 1 second window   (approximately 8 reads per second), with a 150ms overlapping;
- A column discretizing the module of acceleration of each accelerometer, defined after a statistic analysis comparing the data of 5 classes;

The final dataset generated with all the derived features is available for public free use at http://groupware.les.inf.puc-rio.br/har.

4.3   Feature Selection

With the purpose of reducing the use of redundant features and select more informative features in relation to the classes, we used Mark Hall's selection algorithm based on correlation [25]. The algorithm was configured to adopt the "Best First" method, which has a greedy strategy based on backtracking. The 12 features selected by through this procedure were: (1) Sensor on the Belt: discretization of the module of acceleration vector, variance of pitch, and variance of roll; (2) Sensor on the left thigh: module of acceleration vector, discretization, and variance of pitch; (3) Sensor on the right ankle: variance of pitch, and variance of roll; (4) Sensor on the right arm: discretization of the module of acceleration vector; From all sensors: average acceleration and standard deviation of acceleration.

4.4 Classifier for Activity Recognition

Ross Quinlan's [4] C4.5 decision tree was used in connection with the AdaBoost ensemble method [5] for classifying tasks. The C4.5 tree is an evolution proposed by Ross Quinlan to the ID3 algorithm (Iterative Dichotomiser 3) and its main advantage over the ID3 is a more efficient pruning. The boosting AdaBoost method "*tends to generate distributions that concentrate on the harder examples, thus challenging the weak learning algorithm to perform well on these harder parts of the sample space*" [5]. In a simplified manner, with the use of AdaBoost, the C4.5 algorithm was trained with a different distribution of samples in each iteration, thus favoring the "hardest" samples.

We used AdaBoost with 10 iterations and configured the C4.5 tree for a confidence factor of 0.25. The overall recognition performance was of 99.4% (weighted average) using a 10-fold cross validation testing mode, with the following accuracies per class: "sitting" 100%, "sitting down" 96.9%, "standing" 99.8%, "standing up" 96.9%, and "walking" 99.8%. The confusion matrix is presented in Table 3.

Table 3. Confusion Matrix

Sitting	Sitting down	Predicted class Standing	Standing Up	Walking		
50,601	9	0	20	1	Sitting	
10	11,484	29	297	7	Sitting down	
0	4	47,342	11	13	Standing	Actual class
14	351	24	11,940	85	Standing up	
0	8	27	60	43,295	Walking	

The results obtained in this research are very close to the top results of the literature (99.4% in [14], and 99.6% in [15]), even though, it is hard to compare them. Each research used a different dataset, a different set of classes, and different test modes.

5   Conclusion and Future Works

This work discussed a literature review on the recognition of activities using wearable accelerometers data obtained , a wearable device consisted of 4 accelerometers , and the data collection procedure, extraction and selection of features for the development of a classifier for human activities; The main contributions of this article are:

- A comparative table of the researches in the HAR from wearable accelerometers;
- A wearable device for data collection of human activities;
- The offer of a public domain dataset with 165,633 samples and 5 classes, in order to enable other authors to continue the research and compare the results.

In future works we want to include new classes in the dataset and investigate the classifier's performance with the use of accelerometers in different positions and in different quantities. Another future work is the qualitative recognition of activities, which consists in recognizing different specifications for the performance of the same activity, such as different specifications for weight lifting.

Acknowledgements. Wallace Ugulino (142916/2010-2), Debora Cardador (141520/2010-8), and Katia Vega (140859/2010-1) are recipients of individual grants awarded by CNPq. Hugo Fuks is the recipient of individual grants awarded by CNPq (302230/2008-4), FAPERJ/CNE (E-26/102777/2008) and UBILIFE FAPERJ/ADT1-190.116/2010. This work was partially financed by FAPERJ/INC&T (E-26/170028/2008) and CNPq/INCT (557.128/2009-9).

References

1. Yu-Jin, H., Ig-Jae, K., Sang Chul, A., Hyoung-Gon, K.: Activity Recognition Using Wearable Sensors for Elder Care. In: Proceedings of Second International Conference on Future Generation Communication and Networking, FGCN 2008, vol. 2, pp. 302–305 (2008), doi:10.1109/FGCN.2008.165
2. Gjoreski, H., Lustrek, M., Gams, M.: Accelerometer Placement for Posture Recognition and Fall Detection. In: 7th International Conference on Intelligent Environments, IE (2011)
3. Lei, G., Bourke, A.K., Nelson, J.: A system for activity recognition using multi-sensor fusion. In: Annual International Conference of the IEEE Engineering in Medicine and Biology Society, EMBC (2011)
4. Salzberg, S.L.: C4.5: Programs for Machine Learning by J. Ross Quinlan. Morgan Kaufmann Publishers, Inc. (1993) Machine Learning 16(3), 235–240 (1994) ISSN: 0885-6125, doi: 10.1007/bf00993309
5. Freund, Y., Schapire, R.E.: Experiments with a New Boosting Algorithm. In: International Conference on Machine Learning, pp. 148–156 (1996)
6. Liu, S., Gao, R.X., John, D., Staudenmayer, J.W., Freedson, P.S.: Multisensor Data Fusion for Physical Activity Assessment. IEEE Transactions on Biomedical Engineering 59(3), 687–696 (2012) ISSN: 0018-9294
7. Yuting, Z., Markovic, S., Sapir, I., Wagenaar, R.C., Little, T.D.C.: Continuous functional activity monitoring based on wearable tri-axial accelerometer and gyroscope. In: 5th International Conference on Pervasive Computing Technologies for Healthcare, PervasiveHealth (2011)
8. Sazonov, E.S., et al.: Monitoring of Posture Allocations and Activities by a Shoe-Based Wearable Sensor. IEEE Transactions on Biomedical Engineering 58(4), 983–990 (2011)
9. Reiss, A., Stricker, D.: Introducing a modular activity monitoring system. In: Annual International Conference of the IEEE Engineering in Medicine and Biology Society, EMBC (2011)

10. Min, X., Goldfain, A., Chowdhury, A.R., DelloStritto, J.: Towards accelerometry based static posture identification. In: IEEE Consumer Communications and Networking Conference, CCNC (2011)
11. Maekawa, T., Watanabe, S.: Unsupervised Activity Recognition with User's Physical Characteristics Data. In: 15th Annual International Symposium on Wearable Computers, ISWC (2011)
12. Martin, H., Bernardos, A.M., Tarrio, P., Casar, J.R.: Enhancing activity recognition by fusing inertial and biometric information. In: Proceedings of the 14th International Conference on Information Fusion, FUSION (2011)
13. Alvarez-Alvarez, A., Trivino, G., Cordon, O.: Body posture recognition by means of a genetic fuzzy finite state machine. In: IEEE 5th International Workshop on Genetic and Evolutionary Fuzzy Systems, GEFS (2011)
14. Jun-Ki, M., Sung-Bae, C.: Activity recognition based on wearable sensors using selection/fusion hybrid ensemble. In: IEEE International Conference on Systems, Man, and Cybernetics, SMC (2011)
15. Ioana-Iuliana, F., Rodica-Elena, D.: Detection of daily movements from data collected with two tri-axial accelerometers. In: 34th International Conference on Telecommunications and Signal Processing, TSP (2011)
16. Feng, W., Meiling, W., Nan, F.: Research on Classification of Human Daily Activities Based on a Single Tri-Axial Accelerometer. In: 1st International Workshop on Complexity and Data Mining, IWCDM (2011)
17. Czabke, A., Marsch, S., Lueth, T.C.: Accelerometer based real-time activity analysis on a microcontroller. In: 5th International Conference on Pervasive Computing Technologies for Healthcare, PervasiveHealth (2011)
18. Chernbumroong, S., Atkins, A.S., Hongnian, Y.: Activity classification using a single wrist-worn accelerometer. In: 5th International Conference on Software, Knowledge Information, Industrial Management and Applications, SKIMA (2011)
19. Bayati, H., Millan, J.d.R., Chavarriaga, R.: Unsupervised Adaptation to On-body Sensor Displacement in Acceleration-Based Activity Recognition. In: 15th Annual International Symposium on Wearable Computers, ISWC (2011)
20. Atallah, L., et al.: Sensor Positioning for Activity Recognition Using Wearable Accelerometers. IEEE Transactions on Biomedical Circuits and Systems 5(4), 320–329 (2011)
21. Andreu, J., Baruah, R.D., Angelov, P.: Real time recognition of human activities from wearable sensors by evolving classifiers. In: IEEE International Conference on Fuzzy Systems, FUZZ (2011)
22. Xu, S., Kashima, H., Tomioka, R., Ueda, N., Ping, L.: A New Multi-task Learning Method for Personalized Activity Recognition. In: IEEE 11th International Conference on Data Mining, ICDM (2011)
23. Yang, X., Lianwen, J.: A naturalistic 3D acceleration-based activity dataset & benchmark evaluations. In: IEEE International Conference on Systems Man and Cybernetics (SMC), pp. 4081–4085 (2010) ISSN: 1062-922X
24. Maziewski, P., Kupryjanow, A., Kaszuba, K., Czyzewski, A.: Accelerometer signal preprocessing influence on human activity recognition. In: Conference Proceedings of Signal Processing Algorithms, Architectures, Arrangements, and Applications (SPA), pp. 95–99 (2009) ISSN:978-83-62065-06-6
25. Hall, M.A.: Correlation-based Feature Subset Selection for Machine Learning. PhD thesis, Department of Computer Science, University of Waikato, Hamilton, New Zealand (April 1999)

Density-Based Pattern Discovery
in Distributed Time Series

Josenildo C. da Silva[1], Gustavo H.B. Oliveira[1],
Omar A.C. Cortes[1], and Matthias Klusch[2]

[1] Instituto Federal do Maranhão (IFMA)
Depto. Acad. de Informática
65.030-005, São Luís, MA, Brazil
[2] Deutsches Forschungszentrum für Künstiliche Intelligenz (DFKI GmbH)
Forschungsbereich Agenten und Simulierte Realität
Campus Gebäude D 3 2, D-66123 Saarbrücken, Germany

Abstract. Time series data is a very common kind of data in many
different fields. In particular, unknown frequent pattern discovery is one
of the core activities in many time series mining algorithms. Several
solutions to pattern discovery have been proposed so far. However, all
solutions assume centralized dataset. With increasingly development of
network technology distributed data analysis has become popular, raising
issues like scalability and cost minimization. Additionally, some scenarios
such as mining distributed medical or financial data involves the question
of how to preserve data privacy. In this paper, we present a density based
pattern discovery algorithm for time series, which is shown to be efficient
and privacy-preserving.

Keywords: privacy-preservation, time series, pattern discovery.

1 Introduction

Time series data is one of the most common types of data generated in real
world scenarios. Scientific experiments, space telemetry, medical care, financial
and business applications generate an enormous amount of time series data.
Several time series mining algorithms have been proposed in the data mining
literature, such as time series classification or rule discovery [4,12,5]. However,
there are still a number of challenges to be addressed, such as data distribution
and data privacy.

In real world scenarios, datasets are normally split among different sites. Classic approach to mine distributed data is to gather all data in a single central
repository and apply a mining algorithm to find patterns in the centralized data
[14]. Although appealing for most scenarios, this approach may be unfeasible due
to bandwidth usage restrictions. Moreover, commercial or scientific databases are
normally huge, being measured in Terabytes or even larger. Therefore, centralization in such cases is not an option. Distributed data mining addresses these
questions, providing a number of techniques based on ideas such as meta learning

L.N. Barros et al. (Eds.): SBIA 2012, LNAI 7589, pp. 62–71, 2012.

[14], distributed function learning [7], to name a few. Another crucial question in distributed scenarios is how to preserve privacy of sensitive data. In fact, many countries have privacy regulations controlling which data can be disclosed beyond its original purposes, e.g. medical information, or credit card transactions. Privacy-preserving data mining is a research field that address the general problem of providing good data mining results without violating data privacy in the process [19,1]. Many solutions to specific data mining tasks have been proposed, such as association rules [16], classification [6,3], and data clustering (*clustering*) [17], among others.

In this paper, we investigate privacy preserving unknown pattern discovery, since this is one important step in many time series mining algorithms [9,11]. Our approach is based on the assumption that time series patterns can be represented as multidimensional points in a $\mathbb{R}^n$ data space. Therefore, we can reduce the problem of searching for frequent patterns to the search for densely populated regions in this new data space. Our results show that this approach is time and space efficient. Moreover, this approach allows us to cope with privacy requirements since raw data is not transmitted among the parties.

We assume that data is horizontally distributed among a set of parties. Each party holds a mining and a data agent which are responsible for taking part in a mining session to access to local datasets, respectively. Datasets are sensitive and should not be disclosed to other parties. Each party may set a local privacy threshold indicating the minimum amount of privacy it requires to join in a specific mining session. Agents are organized in a pure peer to peer network such that each party may act as initiator or as an arbitrary party in a given mining session. We also assume that agents are semi-honest, i.e. they follow the protocol, but are curious enough to try to discover any sensitive data from other parties whenever possible.

Our main contributions are: (i) we propose DPD-TS, an algorithm based on the idea that frequent patterns discovery can be reduced to finding density regions in an appropriate data space; (ii) we show how privacy is given by this data transformation approach; and (iii) we propose a privacy metric which is suitable for distributed time series data.

The remainder of this paper is organized as follows. Section 2 presents background on our notation and definitions. We discuss privacy metrics in Sec. 3. DPD-FS algorithm and its privacy properties are discussed in Sec. 4. Experiments are presented in Sec. 5. We discuss related work and conclusion in Sec. 6 and 7, respectively.

2   Definitions and Problem Statement

Let $f : \mathbb{N} \to \mathbb{R}$ be a function from time stamps to reals. We define a **time series** $T = \langle x_1, x_2, \ldots, x_m \rangle$, with $x_t = f(t)$ and $1 \leq t < m$, where f is some measurement function. The length of the time series T is denoted by $|T|$. A **subsequence** of T is denoted $\langle x_t, \ldots, x_{t+v} \rangle$, for given integers $1 \leq t < m$ and $1 \leq v < m - t$. Given a time series T with size m and a query subsequence

$Q = \langle q_1, \ldots, q_v \rangle$, $1 \leq q < m$, a **match** is a subsequence $S = \langle s_1 \ldots, s_v \rangle$ of T which satisfies $d(Q, S) \leq r$, for a given distance function d and threshold r, which works as a radius around the a ball centered at Q. A subsequence Q is **reoccurring** if (i) there is more than one match for this query in T; and (ii) for every pair of matching subsequences R and S from T, it holds that R and S do not overlap. A reoccurring subsequence Q is said to be k-**frequent** if Q is in the top-k list of reoccurring subsequences from T. Throughout this paper we use the term **pattern** as a synonym to k-*frequent subsequence*.

Problem 1 (PP-DPDTS). Given a real-valued time series T, an integer k, and a set of sites $\mathcal{L} = \{L_i\}_{1 \leq i \leq P}$, each of them with a local time series T_i, find the set $\mathcal{P}$ of the k-most frequent patterns occurring in $T = \bigcup_{i=1}^{P} T_i$, such that:

1. The total communication cost regarding number of messages and size of each message is minimized;
2. The result using the distributed data T_i is the same if the algorithm runs using $T = \bigcup_{i=1}^{P} T_i$;
3. No party learns about specific values stored in other parties, up to a user defined privacy threshold.

3   Privacy Metric for Distributed Time Series Mining

Privacy can be informally defined as the right to keep some information hidden from other people. Time dimension in data introduces several aspects which may describe a process evolution through a period of time [20], such as *amplitude* (the value of time series at a particular time point), *peaks* (extreme values assumed in the series which may indicate a sudden change of normal behavior), *predictions*, or trends (probable future values for a given model). The basic piece of information to all time series aspects is the amplitude, from which other aspects, like peaks and predictions, can be derived. Therefore, in this work the **privacy goal** is to keep amplitude hidden from other peers. Our model follows the entropy-based metric introduced by Agrawal and Aggarwal [2]. The privacy level of a given point x_t in a time series T, modeled by a random variable X, is given by:

$$PR(x_t) = 2^{h(X)} \tag{1}$$

where $h(X)$ is the differential entropy of X, i.e. $h(X) = -\int f(x)log_2(f(x))dx$, and $f(x)$ is the probability density function of X. The probability density function $f(x)$ represents the knowledge an attacker has about the point x_t. If the attacker has no knowledge, we use the uniform distribution, what give us $PR(X) = 2^{log_2(a)} = a$, the size of the interval from where X is drawn. On the other hand, if we know a better model for x_t we can incorporate it in the privacy level naturally. as:

$$PR(T) = \min\{PR(x_t) \mid t = 0, 1, 2, 3, \ldots, |T|\} \tag{2}$$

Algorithm 1. DPD-TS: initiator	**Algorithm 2.** DPD-TS: Party
Input: k, T_i, n, w, Σ, $\mathcal{L}$, r;	**Input:** k, T_i, n, w, Σ, $\mathcal{L}$, r;
Output: $\mathcal{P}$;	**Output:** $\mathcal{P}$;
At the initiator party L_1 do:	At an arbitrary party L_j do:
1: negotiate(k,n,w, Σ,r);	1: negotiate(k,n,w, Σ,r);
2: $T'_1 \leftarrow$ reduceDim(T_1, n, w);	2: $T'_j \leftarrow$ reduceDim(T_j, n, w);
3: $T''_1 \leftarrow$ discret(T'_1, w, Σ);	3: $T''_j \leftarrow$ discret(T'_j, w, Σ);
4: $\hat{\varphi}_1 \leftarrow$ estimateDensity(T''_1, w, r);	4: $\hat{\varphi}_j \leftarrow$ estimateDensity(T''_j, w, r);
//Cooperative sum	5: **receive** $\hat{\varphi}_{j-1}$ **from** L_{j-1};
5: **send** $\hat{\varphi}_1$ **to** L_2;	// Updating with local density
6: **receive** $\hat{\varphi}_p$ **from** L_p	6: $\hat{\varphi}_j \leftarrow \hat{\varphi}_{j-1} + \hat{\varphi}_j$;
// Global density estimate	// Send to next peer and the last one
7: $\hat{\varphi} \leftarrow \hat{\varphi}_p$	sends to initiator
8: $\mathcal{P} \leftarrow$ getCenters($\hat{\varphi}$,k, r);	7: **send** $\hat{\varphi}_j$ **to** $L_{(j \bmod p)+1}$;
9: **send** $\mathcal{P}$ **to** all agent $L_j \in \mathcal{L}$;	8: **receive** $\mathcal{P}$ **from** L_1;

4 The DPD-TS Algorithm

In this section, we present Distributed Pattern Discovery in Time Series (DPD-TS) algorithm, which is a first step towards a solution to PP-DPDTS problem stated in Sec. 3. DPD-TS exploits the fact that a density estimate can be used to find overcrowded regions in a hyperspace. At each peer a local density estimate is computed and together the peers sum up local densities to produce a global density estimate. With the global density estimate, each local mining agent can perform the discovery step to spot frequent subsequences on the local dataset. The pseudocode for DPD-TS is outlined in Algorithms 1 (initiator) and 2 (arbitrary party). Details are discussed in the following.

DPD-TS computes a set of k-frequent patterns occurring in the union of local time series T_i owned by peers in the mining group $\mathcal{L}$. DPD-TS needs the following parameters: T_i is the local dataset, n is the size used to generate subsequences, w is the number of symbols per string, i.e. the string size, Σ is the alphabet used to generate strings, and $\mathcal{L}$ is the set of peers forming the mining group. The parameter r defines the radius of the density ball to be used in the second step. As output, DPD-TS returns a set $\mathcal{P}$ with the *globally* k-most frequent patterns. DPD-TS assumes that peers agreed upon the parameters values before engaging in a mining session.

Dimension Reduction. Function reduceDim() splits the original time series T_i in various non-overlapping subsequences S of size n. For each subsequence S of T_i, a reduced subsequence $\overline{S}$ with size w is computed. Each point of $\overline{S}$ is the average of $\frac{n}{w}$ points of S. This operation (proposed elsewhere [10]) is known as piecewise aggregate approximation (PAA): The resulting time series T' is a concatenation of all reduced subsequences $\overline{S}$ computed from T. Figure 1 illustrate the dimension reduction of a single subsequence.

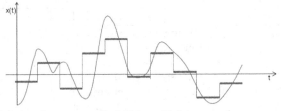
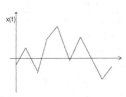
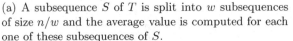

(a) A subsequence S of T is split into w subsequences   (b) Reduced subsequence
of size n/w and the average value is computed for each   is composed of w average
one of these subsequences of S.   values.

Fig. 1. Dimension reduction process

Discretization. Function discret() produces a discretized version of T', which is a
sequence of symbols in a given alphabet. We refer to the discretized version as T''.
For each element x of T', the corresponding string T'' will have a symbol $\sigma_a \in \Sigma$.
The substitution procedure is accomplished by choosing break points $\{\beta_a\}$ in the
values dimension of a given time series T, such that $|\{\beta_a\}| = |\Sigma| - 1$, and such
that each occurrence of a given value x of T'' has the same probability [11],
assuming they are normally distributed. For example, considering a 4-symbol
alphabet, we need 3 break points, where each region will have probability 0.25
of appearing in T. Then, the substitution rule is applied: $x_j = \sigma_a$ if $\beta_{a-1} <
\overline{s_j} \leq \beta_a$, with $1 < a < |\Sigma|$ where $x_j \in T''$.

Estimating Density of Strings. Function estimateDensity() computes the density
estimates of strings S'' from discretized time series T''_j. An important requirement is that the density estimate function $\hat{\varphi}$ builds a non-negative monotonic
function over $\mathbb{R}$ and that the local maxima represent the most dense regions
in the feature space. A general approach to compute data density function
is kernel-based density estimation. For a given kernel function K such that
$\int_{\infty}^{+\infty} K(x)dx = 1$, an estimate of the density, for a specific dataset D, is given
by:

$$\hat{\varphi}[D, r](x) = \frac{1}{Nh} \sum_{x_i \in Neigh(x,r)} K\left(\frac{d(x, x_i)}{h}\right) \tag{3}$$

where N is the total number of points, d is a distance function. The parameter h
is a bandwidth parameter and controls the smoothness of the density estimates[1].
$Neigh(x, r)$ is the set of points close to x, in a given dataset D, which are
inside a ball radius r computed with distance d. We use the triangle kernel
$K(u) = (1 - |u|)I\{|u| \leq 1\}$, where I is the indicator function[2]. We choose this
kernel due to its simplicity, but any other kernel can be used instead. We use

---

[1] Small values of h yields spikier densities and larger values o h yields smother density
   surfaces.
[2] The indicator function, also known as characteristic function, returns 1 if the expression in curly brackets holds and 0 otherwise.

Manhattan distance[3], denoted d, and assume that the alphabet has a total order. An arbitrary point is a string S'' from T''_j. We represent the density estimate as a lookup table holding strings and its density value. Recall that the size of T'' is much smaller than the original time series T. Moreover, each site has only $|T''|/w$ strings to consider, therefore local density is very sparse. This fact is used to minimize communication costs, transmitting only non-zero density entries to other peers.

Computing Global Density. The mining group cooperatively compute the set of globally k-most frequent patterns by summing up all local tables. Initially, peer L_1 sends its local density to L_2. After that, each peer L_j receives partial density estimate $\hat{\varphi}_i$ from its neighbor L_i, with $i = j - 1$ and $j > 1$. L_j adds its own local density to partial global density estimate and sends the partial global density $\hat{\varphi}_j$ to the next neighbor L_{j+1} in the mining group. This protocol continues until the partial sum is sent to L_1, which broadcasts the global density estimate $\hat{\varphi}$ to all members of the mining group. Notice that density estimates are additive, therefore, summing up all local densities produces the same result as estimating a global density from a centralized dataset.

Finding Patterns by Locating Centers. To find the patterns, each peer uses the getCenters(), which works as follows. Choose a set of strings, each of them representing the points that are local maxima in the global density estimate, i.e. centers of the top k most dense regions in the lookup table that stores the density estimates. Let us call the set of k-frequent patterns $\mathcal{P}$. The idea is that local maxima in the pattern space correspond to strings that reoccur more frequently than others do. The regions are constrained to define a ball radius r. More formally, for a given density estimate $\hat{\varphi}$, distance function d, and a string S, we have: $\mathcal{P} = \{S \mid \forall R \in \Sigma^w : (d(S, R) \leq r \to \hat{\varphi}(S) > \hat{\varphi}(R))\}$.

4.1   Performance Analysis of DPD-TS

The time complexity of DPD-TS at a local peer is $O(|T_i|)$. There are $\lfloor \frac{|T_i|}{n} \rfloor$ subsequences in T_i. For each subsequence, w arithmetical means are computed summing $\lfloor \frac{n}{w} \rfloor$ points for each mean, i.e. $O(n)$ steps. Additionally each arithmetical mean is substituted by a symbol, which takes w steps. The overall time cost is $\frac{|T_i|}{n}(n + w) = T_i(\frac{w}{n} + 1)$ steps. Notice that normally $w < n \ll |T_i|$. The discovery step, which is the search for the k-most dense regions, is independent of the size of T_i and is $O(|T|'')$, since we only need to search patterns in the discrete time series T''.

Communication costs are $O(|T''|)$. Each peer sends 1 message to a neighbor peer and receives 1 message from another neighbor. There are only 2 rounds of messages, one of which informs the mining results. Each message has information

---

[3] Manhattan distance, also known as Taxi Cab distance, of two strings is defined as the sum of absolute differences of each character. For example $d(aaa, baa) = 1$, $d(aaa, aab) = 1$ and $d(aaa, bbb) = 3$.

on the density from subsequences found in T_i. For each subsequence there are $2^{|\Sigma|}$ neighbors and there are T_i/w subsequences. However, these values are constants, independent from the size of T''. Therefore, message size is only $O(|T''|)$.

4.2 Privacy Analysis of DPD-TS

A privacy breach occurs when any peer L_i can reconstruct any time series point owned by another peer L_j with arbitrary precision. In the following we analyze how a peer may try to reconstruct information during a DPD-TS session and how much privacy may be enforced.

Initiator peer knows all parameters, the set of global patterns, its own local density estimates and the global density estimates. However, it does not know any local data from other peers. The only information transmitted by the peers during the mining session is the partial density estimate. The initiator has access to the global density estimate. Nevertheless, the density estimate has no information on the order of occurrence of each time series subsequence, which is necessary to reconstruct the original time series. Consequently, an attacker can only try to reconstruct the most provable values for x_t, for a given time stamp t. The initiator also knows the set of global patterns $\mathcal{P}$. However, due to discretization process much information is lost, and therefore cannot be used to reconstruct original data. Indeed, the size of alphabet Σ determines the amount of privacy. The bigger the alphabet, the less privacy we get, because discretized version tends to get the "shape" of the original time series data. Therefore, peers in the mining group can define a minimum amount of privacy they require in order to join the mining session. This is done by setting the maximal size of the alphabet Σ and, consequently, the size of the intervals defined by the breakpoints $\{\beta_j\}$. The privacy level of DPD-TS is defined in Eq. (4).

$$PR(x_t) = 2^{h(x_t)} = 2^{\int_{\beta_j}^{\beta_{j+1}} p(x) log_2 p(x) dx} = 2^{log_2(\beta_{j+1} - \beta_j)} = | \beta_{j+1} - \beta_j | \qquad (4)$$

To improve this reconstruction, attackers need more information, e.g. on the order of occurrence of each pattern. This information, however, is never disclosed during the mining session.

5 Experimental Evaluation

Setup. We implemented DPD-TS in Java 1.6 and run experiments on a Pentium Dual Core 2.33 GHz machine with 2GB memory. For the experiments reported here, we used the following datasets from UCR [8]: *powerdata, sunspot* and *tide* dataset[4]. Results are the averages of 100 runs.

---

[4] All datasets are available at http://www.cs.ucr.edu/ ~eamonn/TSDMA/index.html

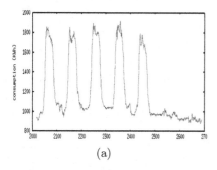

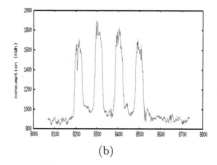

(a)   (b)

Fig. 2. (a) An instance of a *normal week*, the most frequent subsequence in the power data, showing high consumption on workdays and low consumption at weekends. (b) An instance of a *week with holiday* pattern in the power data. In this example, the Monday was a holiday.

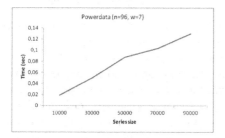

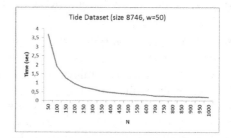

Fig. 3. Time performance (in seconds) as a function of time series size

Fig. 4. Time as a function of n, for tide dataset (8746 points), with $w = 50$

Results. DPD-TS was able to find most frequent pattern in all datasets. For example it found 2 frequent patterns in powerdata (cf. Figure 2). We used $n = 96$, which corresponds to one day (with one measurement every 15 minutes) and $w = 7$, so that we get seven symbols length patterns, representing weeks. We used *Alphabet*= 'abcd' and $r = 1$. The first pattern is "cccccaa" which corresponds to a normal week. The second pattern is "acccaa" which correspond to weeks having a holiday on the first day.

Performance results are shown in Figures 3 and 4. For the performance tests, we created synthetic time series by cloning *powerdata*. The resulting time series have 10 000, 20 000, ..., and 90 000 points. As shown in the Fig. 3, the CPU time increases linearly with the size of the time series. The impact of n in the performance is shown in Fig. 4.

The results of privacy as a function of alphabet size are presented in figure 5. To measure privacy, we used the interval size corresponding to each alphabet symbol in the discretization step, (see Eq. (4)). With just one symbol, the size of interval is given by interval from the minimum to the maximum value observed

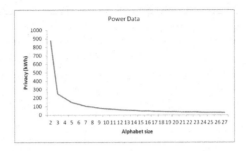

Fig. 5. Privacy level (size of reconstruction interval) with increasing size of alphabet

in the time series, which is 1 538 kWh. With two symbols, the minimum interval is of approximately 3σ, where σ is the standard deviation of the time series[5]. In the figure, we see the decrease of privacy by using more symbols to discretize the time series. With 10 symbols we get a privacy level of 73, 9 kWh. It is up to the user, however, to decide whether or not a given privacy level is enough.

6   Related Work

Privacy-preserving data mining address the general problem of providing useful mining results while providing data privacy. Many solutions have been proposed to specific distributed data mining tasks, e.g. association rules [16], classification [3,13], or data clustering [17]. Research on time series is quite extensive and covers everything from modeling and prediction, to distance metrics and data mining. In the literature one can find several algorithms for finding patterns in time series [18,9,11,12], however, none of these deal with the distribution of data or privacy. More recently, PASTE was proposed as a distributed time series aggregator [15]. PASTE works with the concept of differential privacy, which applies to techniques based on data randomization. Privacy is achieved both through data transformation (Fourier transform), coefficient randomization and cryptographic protocols. DPD-TS by contrast does not use cryptographic methods, although it could be added in next versions. When it comes to privacy metric, DPD-TS follows an information theoretical metric, based on entropy. An accurate comparison of these two metrics is left as future work.

7   Conclusion

We presented DPD-TS, a density-based distributed algorithm to discover unknown patterns in time series data. Results show that this approach is time and space efficient and allows each peer to control how much privacy it needs to preserve. Future work includes extending this approach to multi-variable case and working with variable window width.

---

[5] By the well-known 3 sigma rule for Gaussians.

References

1. Aggarwal, C.C., Yu, P.S. (eds.): Privacy-Preserving Data Mining: Models and Algorithms. Advances in Database Systems, vol. 34. Springer (2008)
2. Agrawal, D., Aggarwal, C.C.: On the design and quantification of privacy preserving data mining algorithms. In: 20th ACM PODS, Santa Barbara, Califonia, pp. 247–255 (May 2001)
3. Agrawal, R., Srikant, R.: Privacy-preserving data mining. In: Proc. of the ACM SIGMOD Conf. on Management of Data, pp. 439–450. ACM Press (May 2000)
4. Davies, P.L., Fried, R., Gather, U.: Robust signal extraction for on-line monitoring data. Journal of Statistical Planning and Inference 122, 65–78 (2004)
5. Elfeky, M.G., Aref, W.G., Elmagarmid, A.K.: Using Convolution to Mine Obscure Periodic Patterns in One Pass. In: Bertino, E., Christodoulakis, S., Plexousakis, D., Christophides, V., Koubarakis, M., Böhm, K. (eds.) EDBT 2004. LNCS, vol. 2992, pp. 605–620. Springer, Heidelberg (2004)
6. Evfimievski, A., Srikant, R., Agrawal, R., Gehrke, J.: Privacy preserving mining of association rules. In: KDD 2002, Edomonton, Alberta, Canada (2002)
7. Kargupta, H., Park, B.-H., Hershberger, D., Johnson, E.: Collective data mining: A new perspective toward distributed data mining. In: Kargupta, H., Chan, P. (eds.) Advances in Distributed and Parallel Knowledge Discovery, ch. 5, part II, pp. 131–174. AAAI Press / MIT Press, Menlo Park, CA / Cambridge, MA (2000)
8. Keogh, E., Folias, T.: The UCR Time Series Data Mining archive. Riverside CA. university of california - computer science & engineering department (2002), http://www.cs.ucr.edu/~eamonn/TSDMA/index.html
9. Keogh, E., Lonardi, S., Chiu, B.: Finding surprising patterns in a time series database in linear time and space. In: KDD 2002, Edmonton, Alberta, Canada, pp. 550–556 (July 2002)
10. Keogh, E.J., Chakrabarti, K., Pazzani, M.J., Mehrotra, S.: Dimensionality reduction for fast similarity search in large time series databases. Knowledge and Information Systems 3(3), 263–286 (2000)
11. Lin, J., Keogh, E., Lonardi, S., Patel, P.: Finding motifs in time series. In: 2nd Works. on Temporal Data Mining, Edmonton, Alberta, Canada (July 2002)
12. Moerchen, F., Ultsch, A.: Discovering temporal knowledge in multivariate time series. In: GfKl 2004, Dortmund, Germany (2004)
13. Pinkas, B.: Cryptographic techniques for privacy-preserving data mining. ACM SIGKDD Explorations 4(2), 12–19 (2002)
14. Prodromidis, A.L., Chan, P.K.: Meta-learning in distributed data mining systems: issues and approaches. In: Kargupta, H., Chan, P. (eds.) Advances in Distributed Data Mining. AAA/MIT Press (2000)
15. Rastogi, V., Nath, S.: Differentially private aggregation of distributed time-series with transformation and encryption. In: Proc. Intl. Conf. on Management of Data 2010, SIGMOD 2010, pp. 735–746. ACM, New York (2010)
16. Saygin, Y., Verykios, V.S., Elmagarmid, A.K.: Privacy preserving association rule mining. In: Reseach Issues in Data Engineering, RIDE (2002)
17. da Silva, J.C., Klusch, M., Lodi, S., Moro, G.L.: Secure agent-based distributed data clustering. Intl. Journal of Web Intelligence and Agent Systems 4(2) (2006)
18. Tanaka, Y., Iwamoto, K., Uehara, K.: Discovery of time-series motif from multi-dimensional data based on mdl principle. Machine Learning 58, 269–300 (2005)
19. Terrovitis, M.: Privacy preservation in the dissemination of location data. SIGKDD Explorations 13(1), 6–18 (2011)
20. Zhu, Y., Fu, Y., Fu, H.: On Privacy in Time Series Data Mining. In: Washio, T., Suzuki, E., Ting, K.M., Inokuchi, A. (eds.) PAKDD 2008. LNCS (LNAI), vol. 5012, pp. 479–493. Springer, Heidelberg (2008)

Filter Approach Feature Selection Methods to Support Multi-label Learning Based on ReliefF and Information Gain

Newton Spolaôr[1,2], Everton Alvares Cherman[1],
Maria Carolina Monard[1], and Huei Diana Lee[2]

[1] Laboratory of Computational Intelligence
Institute of Mathematics and Computer Science
University of São Paulo
São Carlos, Brazil
[2] Laboratory of Bioinformatics
Western Paraná State University
Foz do Iguaçu, Brazil
newtonspolaor@gmail.com, {echerman,mcmonard}@icmc.usp.br,
hueidianalee@gmail.br

Abstract. In multi-label learning, each example in the dataset is associated with a set of labels, and the task of the generated classifier is to predict the label set of unseen examples. Feature selection is an important task in machine learning, which aims to find a small number of features that describes the dataset as well as, or even better, than the original set of features does. This can be achieved by removing irrelevant and/or redundant features according to some importance criterion. Although effective feature selection methods to support classification for single-label data are abound, this is not the case for multi-label data. This work proposes two multi-label feature selection methods which use the filter approach. This approach evaluates statistics of the data independently of any particular classifier. To this end, ReliefF, a single-label feature selection method and an adaptation of the Information Gain measure for multi-label data are used to find the features that should be selected. Both methods were experimentally evaluated in ten benchmark datasets, taking into account the reduction in the number of features as well as the quality of the generated classifiers, showing promising results.

1 Introduction

Machine Learning (ML), which has significant overlapping with data mining, pattern recognition and parts of statistics, is an important field of Artificial Intelligence. ML deals with the fundamental problem of using a dataset to reproduce the process that generated the data.

Multi-label learning deals with the classification problem where each example (or instance) in the training dataset is associated with a set of labels, *i.e.* each example can belong to multiple different classes simultaneously. Multi-label

L.N. Barros et al. (Eds.): SBIA 2012, LNAI 7589, pp. 72–81, 2012.

learning is an emerging research topic due to the increasing number of applications where examples are annotated with more than one label. Multi-label classification has been used in applications such as semantic annotation of video and image, bioinformatics, text categorization and categorization of music into emotions [15].

The task of a multi-label classifier is to predict the label set of unseen examples. Thus, multi-label learning is more general than single-label learning, in which each example in the training dataset is associated with only one class, which can assume several values. Whenever there are more than two class values in single-label learning, it is called multi-class classification. Case the class value is Yes/No, it is called binary classification. In fact, the main difference between multi-label and single-label learning is that classes in multi-label learning are often correlated while the class values in single-label learning are mutually exclusive.

Several approaches have been proposed for multi-label learning, which are well described in [15], where the existing methods for multi-label classification are divided into two main categories: problem transformation and algorithm adaptation. The first category considers methods which transform the multi-label classification problem into either one multi-class classification problem or several binary classification problems. Thus, state of the art algorithms such as SVM can then be used directly. The second category consists of methods that extend specific algorithms such that they can handle multi-label data directly.

Similarly to other data mining and machine learning tasks, multi-label learning also experiences the *curse of dimensionality*, which may cause problems when learning from high-dimensional data. Dimensionality reduction can be tackled, among others, through Feature Selection (FS), which aims to find a small number of features that describes the dataset as well as, or even better, than the original set of features does [8]. This can be achieved by removing irrelevant and/or redundant features according to some importance criterion. Although effective feature selection methods to support classification for single-label data have been extensively studied for many years, few results on multi-label dimensionality reduction have been reported.

This work proposes two multi-label feature selection methods which use the filter approach. This approach evaluates statistics of the data irrespective of any particular classifier. The first method uses the standard approach, which consists in measuring the contribution of each feature according to each label. Afterwards, the average of the score of each feature across all labels is considered, and features with an averaged score greater than a threshold are selected. To this end, ReliefF, a single-label feature selection method is used. Although this approach to multi-label feature selection is standard, to the best of our knowledge this is the first time that ReliefF is used for this purpose. However, this approach does not consider label correlations. The second method uses an adaptation of the Information Gain (IG) measure for multi-label data, and features which have an IG greater than a threshold are the ones selected.

Both methods were experimentally evaluated in ten benchmark datasets, taking into account the reduction in the number of features as well as the quality of the generated multi-label classifiers, showing promising results.

The rest of this paper is organized as follows: Section 2 briefly presents multi-label learning and Section 3 addresses feature selection for multi-label learning as well as related work. The filter methods proposed are described in Section 4 and their experimental evaluation in Section 5, which is followed by the conclusions and future work in Section 6.

2   Multi-label Classification

This section presents basic concepts and terminology of multi-label learning, as well as the Binary Relevance multi-label transformation approach used in this work.

2.1   Basic Terminology and Concepts

Let D be a dataset composed of N examples $E_i = (\mathbf{x}_i, Y_i)$, $i = 1..N$. Each example E_i is associated with a feature vector $\mathbf{x}_i = (x_{i1}, x_{i2}, \ldots, x_{iM})$ described by M features X_j, $j = 1..M$, and a subset of labels $Y_i \subseteq L$, where $L = \{y_1, y_2, ...y_q\}$ is the set of q possible labels. Table 1 shows this representation. In this scenario, the multi-label classification task consists in generating a classifier H which, given an unknown instance $E = (\mathbf{x}, ?)$, is capable of accurately predicting its subset of labels Y, i.e., $H(E) \to Y$.

Table 1. Multi-label data

	X_1	X_2	...	X_M	Y
E_1	x_{11}	x_{12}	...	x_{1M}	Y_1
E_2	x_{21}	x_{22}	...	x_{2M}	Y_2
$\vdots$	$\vdots$	$\vdots$	$\ddots$	$\vdots$	$\vdots$
E_N	x_{N1}	x_{N2}	...	x_{NM}	Y_N

As already mentioned, methods for multi-label classification can be divided into two main categories: problem transformation and problem adaptation. The first category considers methods which transform the multi-label classification problem into either one multi-class classification, such as the Label Powerset (LP) approach, in which each unique set of labels in the training set is considered as a class value, or several binary classification problems, such as the Binary Relevance approach described next does. In both cases, multi-class or binary, respectively, state of the art algorithms can then be used directly. The second category consists of methods that extend specific algorithms such that they can handle multi-label data directly [15].

2.2   Binary Relevance

This problem transformation approach decomposes a multi-label classification problem into several distinct binary classification problems, one for each label in the set of labels L with $|L| = q$. The Binary Relevance (BR) approach initially transforms the original training dataset into q binary datasets D_{y_j}, $j = 1..q$, where each D_{y_j} contains all examples of the original dataset, but with a single positive or negative label related to the single label y_j according to the true label subset associated with the example, *i.e.*, positive if the label set contains label y_j and negative otherwise. The other labels $(y_k, k \neq j)$ are not included in D_{y_j}. After the data is transformed, a set of q binary classifiers $H_j(E), j = 1..q$ is constructed using the correspondent training dataset D_{y_j}. In other words, the BR approach initially constructs a set of q classifiers — Equation 1:

$$H_{BR} = \{C_{y_j}((\mathbf{x}, y_j)) \rightarrow \lambda_j \in \{0, 1\} | y_j \in L : j = 1..q\} \tag{1}$$

To classify a new multi-label instance, the algorithm outputs the aggregation of the labels positively predicted by all the q independent binary classifiers.

An advantage of the BR approach is its low computational complexity compared with other multi-label methods. For a constant number of examples, BR scales linearly with size q of the label set L, which makes it appropriate for not very large q. For large numbers of labels some divide-and-conquer methods have been proposed to organize labels into a tree-shaped hierarchy where it is possible to deal with a much smaller set of labels compared to q. A disadvantage of the standard BR approach is that it completely ignores any label relationships. However, two successful methods that enable the binary classifiers to discover existing label dependency by themselves have already been proposed [2,10].

3   Feature Selection for Multi-label Classification

Approaches to feature selection are addressed next, as well as related work in feature selection for multi-label classification.

3.1   Feature Selection Approaches

FS methods can be classified into three main categories (wrapper, embedded or filter) according to the interaction with the learning algorithm [8].

The wrapper approach uses the learning algorithm itself as a black box to evaluate candidate subsets of features, repeating the process on each feature subset until a stopping criterion is met. Thus, wrapper methods take into consideration all the important characteristics of the learning algorithm in the final decision of the feature selection process. However, its computational cost could be very high. Similarly to wrappers, FS performed by embedded methods is linked with the learning algorithm itself. However, in this case this link is stronger than in wrappers, since the FS process is included in the classifier construction. A typical example of embedded methods for feature subset selection is decision trees [9]. Unlike these two approaches, filter methods perform a separate process that does not interact with and is independent from the learning algorithm itself. The basic idea of filters is to use general characteristics of data to select the relevant features according to these characteristics, before the construction of the classifier takes place. An advantage of filters is the fact that they are fast and simple to use.

3.2   Related Work

Although effective FS methods to support classification for single-label data have been extensively studied for many years, few results on multi-label dimensionality reduction have been reported. A systematic review process, a method to support bibliographic reviews, related to multi-label FS was carried out in [12]. Results show the findings of less than 50 related papers, as well as a growing interest in the subject in recent years.

Some papers use the wrapper approach addressing directly the multi-label data [17]. However, most papers consider the previous transformation of multi-label data to multi-class data (using LP) or binary data (using BR). Afterwards, the filter approach is used in the transformed data. To this end, measures related to Information Gain [1,16], mutual information [5], chi-square [14] and others are used. Whenever the BR approach is used, each label is considered separately and the results are combined using, for example, an averaging approach. Embedded feature selection is used in [3,6]. In addition, in [7] it is proposed to learn the label correlation and do FS simultaneously.

4   Multi-label Feature Selection Methods Proposed

The first method, named RF, was initially proposed in [13], where it was evaluated but on few multi-label datasets. RF uses ReliefF, an algorithm which measures the quality of attributes of single-label data. The main advantage of ReliefF over other strictly univariate measures is that it takes into account the effect of interacting attributes. The idea of ReliefF and its derivatives is to reward an attribute for having different values on a pair of nearest examples from different classes, and penalize it for having different values on examples from the same class [4,11]. For each feature, ReliefF outputs a value w, ranging from -1 to 1 with large positive w assigned to important features.

Initially, RF uses the BR approach to transform the multi-label training dataset into q binary datasets and ReliefF is used in the conventional way to evaluate the set of features $\{X_1, X_2, ..., X_M\}$ on each of the q binary datasets. The q ReliefF measure values of each feature $X_i, i = 1..M$, are then averaged and the ones with values greater than or equal to a threshold are selected. However, apart from the use of ReliefF, of which we are not aware it has been used before for multi-label feature selection, RF uses the standard multi-label filter approach, which considers each label separately. Thus, it has the disadvantage of do not considering label correlations.

The second method, named IG-ML, aims at taking into consideration label correlations. To this end, the Information Gain (IG) measure for multi-label data proposed in [3] is used directly in the multi-label data. IG-ML evaluates the multi-label IG of the set of features $\{X_1, X_2, ..., X_M\}$ and the ones with IG values greater than or equal to a threshold are selected.

5   Experimental Evaluation

Both methods were implemented using Mulan[1], a Java package for multi-label classification based on Weka[2]. The experiments were carried out using two different base single-label classifiers from Weka: J48, an implementation of the decision tree C4.5 algorithm [9] and the support vector machine SMO learning algorithm. The methods were evaluated using 10 datasets. All the reported results were obtained by Mulan using 10-fold cross validation with paired folds.

5.1   Datasets

Table 2 describes the datasets used in the experiments, obtained from the Mulan's repository[3]. It shows the datasets domain (Domain); number of examples (N); number of features (M); number of labels ($|L|$); Label Cardinality (LC), which is the average number of labels associated with each example defined by Equation 2; Label Density (LD), which is the normalized cardinality defined by Equation 3, and the number of Distinct Combinations (DC) of labels.

$$LC(D) = \frac{1}{|D|} \sum_{i=1}^{|D|} |Y_i| \qquad (2) \qquad LD(D) = \frac{1}{|D|} \sum_{i=1}^{|D|} \frac{|Y_i|}{|L|} \qquad (3)$$

Table 2. Datasets used for experiments

Dataset	Domain	N	M	|L|	LC	LD	DC
1-*bibtex*	text	7395	1836	159	2.40	0.02	2856
2-*cal500*	music	502	68	174	26.04	0.15	502
3-*corel16k001*	images	13766	500	153	2.86	0.019	4803
4-*corel5k*	images	5000	499	374	3.52	0.01	3175
5-*emotions*	music	593	72	6	1.87	0.31	27
6-*enron*	text	1702	1001	53	3.38	0.06	753
7-*genbase*	biology	662	1186	27	1.25	0.05	32
8-*medical*	text	978	1449	45	1.25	0.03	94
9-*scene*	image	2407	294	6	1.07	0.18	15
10-*yeast*	biology	2417	103	14	4.24	0.30	198

5.2   Performance Measures

The performance of multi-label classifiers can be evaluated using different measures. Some of these measures are adaptations from the single-label classification

---

[1] http://mulan.sourceforge.net
[2] http://www.cs.waikato.ac.nz/ml/weka/
[3] http://mulan.sourceforge.net/datasets.html

problem, while others were specifically defined for multi-label tasks. In what follows, we briefly describe the measures used in this work to compare both methods. These measures are *Hamming Loss*, *Accuracy*, *F-Measure* and *Subset Accuracy*, defined by Equations 4 to 7 respectively, where Δ represents the symmetric difference between two sets, Y_i is the set of true labels, Z_i is the set of predicted labels and $I(\text{true}) = 1$ and $I(\text{false}) = 0$.

$$Hamming\,Loss(H, D) = \frac{1}{N} \sum_{i=1}^{N} \frac{|Y_i \Delta Z_i|}{|L|} \tag{4}$$

$$Accuracy(H, D) = \frac{1}{N} \sum_{i=1}^{N} \frac{|Y_i \cap Z_i|}{|Y_i \cup Z_i|} \tag{5}$$

$$F\text{-}Measure(H, D) = \frac{1}{N} \sum_{i=1}^{N} \frac{2|Y_i \cap Z_i|}{|Z_i| + |Y_i|} \tag{6}$$

$$Subset\,Accuracy(H, D) = \frac{1}{N} \sum_{i=1}^{N} I(Z_i = Y_i) \tag{7}$$

All these performance measures have values in the interval $[0..1]$. For *Hamming Loss*, the smaller the value, the better the algorithm performance is, while for the other measures greater values indicate better performance. Note that *Subset Accuracy* is a very strict evaluation measure as it requires an exact match of the predicted and the true set of labels. Furthermore, as one of the advantages of FS is to reduce the data dimensionality, the average percentage of feature reduction in RF and IG-ML was also considered in the experimental evaluation.

5.3   Results and Discussion

Table 3 presents the average feature reduction and the standard deviation (in brackets) carried out by RF and IG-ML, using as a threshold 0.01 and 0.1 respectively, which can be considered conservative [11].

Table 3. Average percent of feature reduction (and standard deviation)

Dataset	RF	IG-ML
1-*bibtex*	78.31(0.31)	84.79(0.30)
2-*cal500*	8.82(0.98)	0.00(0.00)
3-*corel16k001*	70.10(0.67)	–
4-*corel5k*	43.99(1.62)	95.93(0.31)
5-*emotions*	23.89(1.94)	1.53(0.44)
6-*enron*	1.27(0.30)	6.22 (0.54)
7-*genbase*	95.51(0.21)	97.05 (0.07)
8-*medical*	86.62(1.06)	95.89 (0.05)
9-*scene*	19.15(0.58)	11.63 (1.64)
10-*yeast*	40.58(2.74)	15.73 (2.85)

Note that the IG values of all features for dataset *corel16k001* were lower than the threshold (8 cases). As can be observed, for some datasets both methods reduced the number.of features by more than 75% (*bibtex*, *genbase* and *medical* datasets). For *cal500* RF reduced less than 10% while IG-ML no reduced dimensionality. However, for *corel5k*, IG-ML was able to reduce twice as much features than RF. Thus, we can conclude that, in general, RF and IG-ML select different features. Nevertheless, the features selected must be useful for the multi-label learning algorithm. To this end, using the BR approach and the two different base-learning algorithms J48 and SMO, the classifiers constructed using all features, and the features selected by RF and IG-ML were analyzed.

For all datasets and base-learning algorithms, the average (and the standard deviation) of the four multi-label performance measures were tabulated. Observe that the value of these measures for the classifiers constructed using all features represent a good *Baseline* for the ones obtained with the features selected by RF and IG-ML. However, due to lack of space, these tabulated results are not shown in this paper, but they can be found at `http://www.labic.icmc.usp.br/pub/mcmonard/ExperimentalResults/SBIA2012.pdf`.

These results show that from the total of 152 performance measure values tabulated (2 base-learning algorithms × 2 FS methods × 4 performance measures × 10 datasets −8), and considering the standard deviation (which has 0.07 as its maximum value), only 21 of them show a degradation compared to the correspondent *Baseline* performance measure. This represents less than 14%, which can be considered a good result. Furthermore, 9 of these 21 cases were obtained when using J48, and the remaining 12 by SMO. Regarding the datasets, most of the performance measures were significantly worse than those for the correspondent *Baseline* for *corel5k* (10 cases, 5 by RF and 5 by IG-ML). From the remaining 11 cases, 10 were obtained using the features selected by RF in the following datasets using the specified base-learning algorithm (this behavior happens to both algorithms case the learning algorithm is not specified), and performance measures: 1-*bibtex* (*Subset Accuracy* & SMO, *F-Measure* and *Accuracy*); 3-*corel16k001* (*F-Measure* and *Accuracy*, *Subset Accuracy* & J48). The last case was obtained by IG-ML in the *bibtex* dataset (*Hamming Loss* & SMO).

Nevertheless, in order to evaluate FS two aspects should be considered simultaneously: the reduction in the number of features *versus* the performance measure values of the classifier generated using the features selected. To this end, a graphical analysis is more appropriate. It should be observed that the evaluation of multi-label learning is more difficult than for single-label learning. In fact, the classification of a new instance by single-label classifiers has only two possible outcomes: correct or incorrect. On the other hand, multi-label classifiers should also take into account partially correct classification. Thus, several performance measures have to be analyzed. Due to lack of space, only one graph with this kind of analysis is shown. Graphs for all datasets and the four performance measures used in this work can be found at `http://www.labic.icmc.usp.br/pub/mcmonard/ExperimentalResults/SBIA2012.pdf`.

To illustrate, Figure 1 shows this information for *Hamming Loss* and *F-Measure* using *bibtex* dataset. *Baseline* refers to the performance measure values obtained using all features. One can note that for *Hamming Loss*, results nearer to the left-hand bottom corner of the figure are the best, while for *F-Measure* the best results are the ones nearer to the right-hand bottom corner.

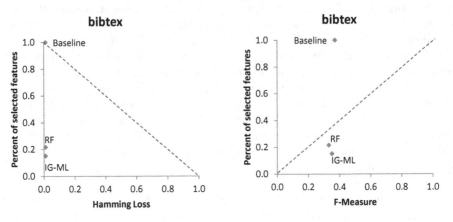

Fig. 1. *Bibtex* dataset FS evaluation using J48 as base-learning algorithm

As can be observed in Figure 1, both FS methods obtained good results, since the performance measure values are similar to the *Baseline*, but were obtained using less features. Moreover, for this dataset, IG-ML is slightly better than RF since it was able to reduce the number of features more. Similar results for this dataset were obtained using SMO as base-learning algorithm.

6   Conclusion

Selecting features is an important task in machine learning in order to take care of the *curse of dimensionality* problem. This work analyses the behavior of two feature selection methods for multi-label learning which use the filter approach. The first method, RF, uses the multi-label feature selection standard approach. This approach considers a feature evaluation measure for each label separately, which are further composed in only one evaluation measure used to select the features. In this work, we proposed the use of ReliefF to evaluate each feature separately. However, the standard approach fails to consider any correlation among labels. The second method, IG-ML, aims to take into account the correlation among labels. To this end, the new information gain measure proposed in the literature for multi-label learning is used directly in the multi-label data as a feature evaluation measure. Both FS methods were thoroughly evaluated experimentally in ten benchmark datasets, showing promising results.

As future work, we plan to investigate the possibility of extending the ideas behind ReliefF for multi-label data.

Acknowledgment. The authors would like to thank the anonymous referees for their insightful comments on this paper. This research was supported by the Brazilian Research Council FAPESP.

References

1. Chen, W., Yan, J., Zhang, B., Chen, Z., Yang, Q.: Document transformation for multi-label feature selection in text categorization. In: IEEE International Conference on Data Mining, pp. 451–456 (2007)
2. Cherman, E.A., Metz, J., Monard, M.C.: Incorporating label dependency into the binary relevance framework for multi-label classification. Expert Systems with Applications 39(2), 1647–1655 (2012)
3. Clare, A., King, R.D.: Knowledge Discovery in Multi-label Phenotype Data. In: Siebes, A., De Raedt, L. (eds.) PKDD 2001. LNCS (LNAI), vol. 2168, pp. 42–53. Springer, Heidelberg (2001)
4. Demšar, J.: Algorithms for subsetting attribute values with Relief. Machine Learning 78, 421–428 (2010)
5. Doquire, G., Verleysen, M.: Feature Selection for Multi-label Classification Problems. In: Cabestany, J., Rojas, I., Joya, G. (eds.) IWANN 2011, Part I. LNCS, vol. 6691, pp. 9–16. Springer, Heidelberg (2011)
6. Esuli, A., Fagni, T., Sebastiani, F.: Boosting multi-label hierarchical text categorization. Information Retrieval 11(4), 287–313 (2008)
7. Gu, Q., Li, Z., Han, J.: Correlated multi-label feature selection. In: ACM International Conference on Information and Knowledge Management, pp. 1087–1096 (2011)
8. Liu, H., Motoda, H.: Computational Methods of Feature Selection. Chapman & Hall/CRC (2008)
9. Quinlan, J.: C4.5: Programs for Machine Learning. Morgan Kaufmann (1993)
10. Read, J., Pfahringer, B., Holmes, G., Frank, E.: Classifier Chains for Multi-label Classification. In: Buntine, W., Grobelnik, M., Mladenić, D., Shawe-Taylor, J. (eds.) ECML PKDD 2009, Part II. LNCS, vol. 5782, pp. 254–269. Springer, Heidelberg (2009)
11. Robnik-Sikonja, M., Kononenko, I.: Theoretical and empirical analysis of ReliefF and RReliefF. Machine Learning 53(1-2), 23–69 (2003)
12. Spolaôr, N., Monard, M.C., Lee, H.D.: A systematic review to identify feature selection publications in multi-labeled data. ICMC Technical Report No. 374, University of São Paulo, p. 31 (2012)
13. Spolaôr, N., Cherman, E.A., Monard, M.C.: Using ReliefF for multi-label feature selection. In: Conferencia Latinoamericana de Informática, pp. 960–975 (2011) (in portuguese)
14. Trohidis, K., Tsoumakas, G., Kalliris, G., Vlahavas, I.: Multi-label classification of music into emotions. In: International Conference on Music Information Retrieval, pp. 1–6 (2008)
15. Tsoumakas, G., Katakis, I., Vlahavas, I.: Mining multi-label data. In: Data Mining and Knowledge Discovery Handbook, pp. 1–19 (2009)
16. Wei, Q., Yang, Z., Junping, Z., Wang, Y.: Semi-supervised multi-label learning algorithm using dependency among labels. In: International Conference on Machine Learning and Computing, pp. 112–116 (2009)
17. Zhang, M.L., Peña, J.M., Robles, V.: Feature selection for multi-label naive bayes classification. Information Sciences 179, 3218–3229 (2009)

Automatic Analysis of Leishmania Infected Microscopy Images via Gaussian Mixture Models

Pedro A. Nogueira and Luís Filipe Teófilo

Laboratória de Inteligência Artificial e Ciência de Computadores
Faculdade de Engenharia da Universidade do Porto, FEUP
Porto, Portugal
pedro.alves.nogueira@fe.up.pt

Abstract. This work addresses the issue of automatic organic component detection and segmentation in confocal microscopy images. The proposed method performs cellular/parasitic identification through adaptive segmentation using a two-level Otsu's Method. Segmented regions are divided using a rule-based classifier modeled on a decreasing harmonic function and a Support Vector Machine trained with features extracted from several Gaussian mixture models of the segmented regions. Results indicate the proposed method is able to count cells and parasites with accuracies above 90%, as well as perform individual cell/parasite detection in multiple nucleic regions with approximately 85% accuracy. Runtime measures indicate the proposed method is also adequate for real-time usage.

1 Introduction

Leishmania is the parasite responsible for Leishmaniasis, a disease currently affecting over 12 million people throughout 88 countries [1]. Leishmaniasis is treatable by chemotherapeutics, which, nevertheless, suffer from poor administration regimens and high host toxicity [2]. Although the disease is not generally deadly, it severely damages the immune system, leaving the body exposed to other deadly pathogens, which often prove fatal [3]. The inadequate means to treat Leishmaniasis render the research for new treatments an urgent task.

Research in microscopy imaging produces large amounts of data, which requires anywhere from full days to weeks to classify and annotate. In a single laboratory the number can easily ascend to thousands of images with merely a dozen different experiments. Not only does this detract the researchers from exploring new alternatives, as it also introduces inter-person variance, as many images are extremely cluttered and contain several hundreds of cells and parasites. This results in a time consuming and mentally straining process, which expresses itself as a decaying function over time as the subject gets tired, frustrated or bored.

These reasons justify the need for the development of automatic mechanisms that are able to replace or aid researchers in the annotation task, for which and to the best of our knowledge no current solution exists. The proposed method provides a fully automatic, real-time pipeline for the identification of cells and parasites in Leishmania infected microscopy imaging, thus enabling more accurate annotations.

L.N. Barros et al. (Eds.): SBIA 2012, LNAI 7589, pp. 82–91, 2012.

Pertaining this paper's organization, it is structured as follows. Section 2 describes the main characteristics of fluorescence microscopy imaging, as well as the dataset used in this study. Section 3 discusses the state of the art in cell identification and segmentation in microscopy imaging. Section 4 briefly describes the proposed method, followed by the description of its steps. In section 5, the results for the segmentation step and the implemented classifiers are presented. Finally, section 6, presents conclusions on the developed work, commenting on its performance and readiness for real-world applicability.

2   Fluorescence Microscopy Imaging

In contrast to the classical optical microscopy, the use of fluorescence microscopy allows simultaneous labeling of different cell components, which can be easily distinguished based on the fluorescence properties of their specific dyes [4]. The images collected for this study used three fluorophores, which emitted three distinct wavelengths [4]. These corresponded to the cell nuclei DNA (in blue), cytoplasmic and nuclear DNA (in red) and the parasitic DNA (in green). This provided three separate sets of data per image (Figure 1), motivating the identification of cells, parasites and cytoplasm individually in the three channels as independent images.

Although very popular, fluorescence microscopy imaging (FMI) presents some well-known issues that also characterized our dataset. The most noticeable issues included: non-linear illumination (due to poor lighting conditions and sub-optimal experimental setup), photo bleaching [5], varying contrast, Gaussian noise, chromatic aberrations and overlapping cells, as well as parasites (due to various focal planes).

In this study 794 fluorescence microscopy images from random drug trials with different experimental setups were collected and used. These images were collected through a light microscope and annotated manually by a Leishmania research team at the INEB/IBMC laboratory. Refer to section 6 for further details.

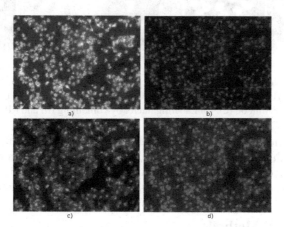

Fig. 1. Details of a fluorescence microscopy image. a) Original image; b) Cell nuclei channel; c) Parasite nuclei channel; d) Cytoplasmic channel.

3   Related Work

Microscopy image analysis has been an active field for several decades. In related work, Liao et al. [6] used a simple thresholding method, coupled with mathematical morphology and contextual shape detection to detect white blood cells. However, their approach does not tolerate cells outside the defined conditions (e.g.: poorly segmented regions, forming a cell cluster region). An automated method for cellular membrane segmentation is described in [7]. This method also allows the reconstruction of un-stained tracts through the nuclear membranes as a spatial reference. Jiang et al. have also proposed white blood cell segmentation using scale-space filtering and watershed clustering in HSV color space [8].

Park has proposed bone marrow cell segmentation through an iteratively relaxed watershed algorithm [9]. However, this work is sensitive to illumination and noise conditions, since it overly relies on the fixed mean color values of each patch for the relaxation procedure. Begelman [10] performs cell nuclei segmentation using color, shape features and a fuzzy logic engine. This work is more robust than the aforementioned one because the extracted shape features, serve as an auxiliary classification input. However, it is still not able to account for non-circular geometries or abnormally colored cells due to the implemented rules' simplicity.

Yu proposes using an adaptive thresholding technique to detect cell nuclei, which are then expanded via level sets to determine cell boundaries [11]. Yan proposes a similar approach [12]. Yan improves on Yu by replacing the adaptive thresholding with a distance map of the initial adaptive histogram-based thresholding step. This distance map is then used to create a watershed transform, serving as a region list representing the level-set seed points (Figure 2). The only drawback to this approach is that it is not able to deal with highly cluttered images, as the distance map would not provide enough information to accurately parameterize the watershed transform, thus leading to an erroneous number/location of seed points for the level-set step.

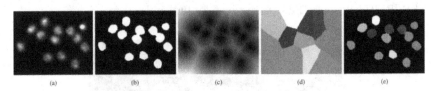

Fig. 2. Segmentation example of the method proposed in [12]. (a) Patch taken from original DNA image. (b) Binary thresholding result of (a). (c) Distance transform of (b); (d) Result of the watershed algorithm on (c). (e) Labeling nuclei by combining (b) and (d).

From this review, it is clear that most of the literature does not attempt or is unable to deal with highly cluttered or overlapped image regions. This is a major concern in microscopy image analysis as the great majority of real-world data is heavily cluttered and saturated. The proposed method in this paper aims at addressing this issue.

4   Proposed Method

The proposed method focuses on developing robust methods for identifying and segmenting cellular and parasitic agglomerations in confocal microscopy images.

The method starts by splitting the original image f in three channels: f_c (blue); f_p (green); f_{cyt} (red). Each channel is then normalized and segmented through Otsu's Method [13]. This yields three region vectors, corresponding each one to cellular DNA, parasitic DNA and weak[1] DNA signatures (cytoplasm). Low-level features are computed for the first two aforementioned region vectors, and then used to train a rule-based classifier and a Support Vector Machine, both of which attempt to classify each region as to how many cells or parasites it contains. To resolve disputes between these two classifiers, a voting system taking into account both of the classifiers' error margins is employed. Each region is then further segmented into the predicted number of sub-regions by Gaussian unmixing [17]. Figure 3 can be inspected bellow for a more structured understanding of the described pipeline.

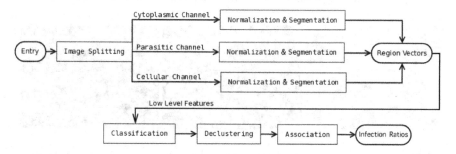

Fig. 3. Developed method's architecture

4.1 Pre-processing and Segmentation

The method first splits each of the target image's f color channels, as they are independently processed. Each image channel is then normalized and segmented into background and foreground components. An initial study on the general image characteristics was conducted in order to choose an appropriate segmentation technique. In this study, the intensity values of 120 randomly selected images presented clear bimodal distributions for all color channels (Figure 4).

Fig. 4. Bimodal distributions observed in the RGB components of 120 images (averaged). A) Red color component; b) Green color component; c) Blue color component.

---

[1] In order to register a weak DNA signature, this fluorophore must be highly sensitive. Thus, it also registers the cell nuclei's DNA. Since the cell nuclei can be trivially subtracted through set operations involving the cell channel, we denominated this channel as the cytoplasmic channel.

Thus, Otsu's Method presented itself as a fitting approach due to its low temporal and spatial complexity, non-parameterisable characteristics and segmentation principle. Otsu's Method's principle assumes a bi-modal distribution in the target dataset, for which it attempts to determine an optimal threshold value t by minimizing intra-class variance (σ) [13]:

$$\sigma_b^2(t) = \sigma^2 - \sigma_\omega^2(t) = \omega_1(t)\, \omega_2(t)\, [\mu_1(t) - \mu_2(t)]^2 \tag{1}$$

Each color channel was binarised using Otsu's Method and then proceeded to a connected component analysis [18], resulting in a region vector representing the cellular, parasitic and cytoplasmic regions present in the image. Note that the cytoplasmic regions are not used in this work, as they are intended for associating cell-parasite pairs. They were, however, computed and integrated into the method in hindsight of future work. Figure 5 depicts the result of the segmentation step for the cellular channel.

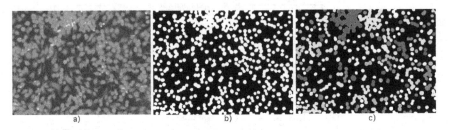

Fig. 5. Segmentation output of a moderately cluttered image. a) Original image; b) Segmentation output; c) Visual representation of the cellular region vector obtained from the connected component analysis (randomly color-coded).

Following the region extraction, various low-level features were computed for each region. From these we highlight the ones with greatest classification potential, which were later chosen to form the following feature vector $F_i = [a,\ ll_{1..N},\ d^1{}_{1..N}(ll),\ d^2{}_{1..N}(ll)]$. The features comprising the vector are: a, the area value (in pixels); ll, the log-likelihood ratios for modeling the region with 1 to N Gaussian mixtures; $d^1(ll)$, the first discrete derivate of ll and $d^2(ll)$, the second discrete derivate of ll. The area feature was used to define the rule-based classifier. The log-likelihood ratios and their derivates were used to train the SVM classifier.

4.2   Classification

The classification step is based on the assumption that the regions obtained from the segmentation process may not always correspond to a single cell or parasite. Based on this, it attempts to discern in how many sub-regions each region must be split. This is achieved by employing two separate classifiers: a rule-based classifier (RBC) and a support vector machine (SVM) [20].

The RBC exploits the low area overlap percentage observed between cell and parasite pairs. This low overlap percentage is due to the high depth of field observed in the collected images, resulting in a near-perfect 2D cross-section of the 3D space within the tissue sample. Following this principle, and as single cells/parasites presented normal distributions, it was hypothesized that the area functions for cells and parasites could be approximated by a harmonic function. In fact, this was verified experimentally with the addition of larger multi-nucleic regions being less frequent. Thus, the functions present decreasing harmonic patterns. Although simple, this approach was found to be quite accurate (around 89.0% accuracy, hitting a maximum value of 98,9% when considering an error of ±2 regions as acceptable). Figure 6 illustrates a detail for one of the observed decreasing harmonic functions.

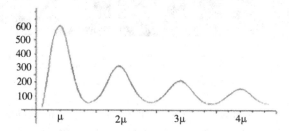

Fig. 6. Detail of a decreasing harmonic function described by the normalized (cellular) area values of 120 randomly sampled images (~870 regions per image: 106.318 region samples total). Horizontal axis: area values; Vertical axis: area value occurrences.

In light of this finding, the classifier was programmed with rules reflecting these functions and taught to ignore values outside its knowledge space, as there was no data to accurately model regions over 10 cells or parasites. Note that cells and parasites each have their own harmonic function, since their area distributions do not exhibit the same standard deviation.

The SVM classifier relies on the concept that circles and ellipses can be described as Gaussian distributions and, as macrophages and parasites partake such geometry, clusters of these objects can be formulated as a mixture problem. Our conjectured hypothesis was that: as mixtures are added to the modeling process of each region, the improvement rate is described in the log-likelihood ratio evolution sequence. Thus, if an initial annotated dataset with the correct number of mixtures N is available, it should be possible to model a function that is able to predict this N for new, non-annotated observations. Following this hypothesis, the classifier was trained with a subset of features from the main feature vector, consisting of the log-likelihood ratios, it's first and second order discrete derivates. The training set was obtained from roughly 150 regions, modeled with $N = [1..15]$ Gaussian mixtures. Various machine-learning classifiers were tested. Ultimately, a SVM model was chosen as it achieved the highest (85,3%) sequential split and cross-validation classification rates.

4.3   Declustering and Association

Having obtained a prediction for the number of nuclei in each region, these were un-mixed using the Expectation-Maximization [14] method. The algorithm was parameterized with a minimum standard deviation of $1x10^{-6}$, and a maximum of 200 iterations. To minimize runtime, the seeds for the centroids of each mixture were set by the averaged centroids of a 10 fold cross-validated K-Means. Figure 7, portrays the meth-od's expert performance, even in the presence of large nucleic clusters.

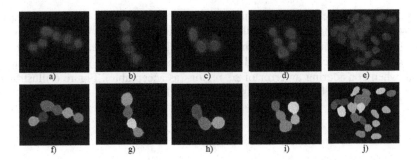

Fig. 7. Several *declustering* examples for 6, 4, 3, 5 and 11 region clusters. a) through e) Original region patches. f) through j) Respective *declustered* patches.

Since not all biomedical researchers use the same annotation guidelines, two routines representing the two most popular techniques [16] were implemented:

- If the cell overlaps with one or more parasites;
- If a parasite is within a certain radius (~50% of the average cell radius).

Cellular and parasitic cytoplasmic membership was computed via set algebra on the region vectors. The radius was calculated using the Euclidean between the centers of mass and boundaries of each declustered region.

5   Results

To appraise the proposed method, the segmentation and classification steps were individually assessed. For this, two stress datasets (DSA, B) were built. Dataset A tested the segmentation step by presenting dim and out-of-focus conditions, while dataset B included highly clustering images, thus straining the classification step. Three biomedical researchers provided the ground truth by manually annotating both datasets. Researchers were asked to carefully perform the annotations in separate days and double-check them, so as to minimize human error.

5.1   Segmentation

Since the method assumes that the cell/parasite identification process is not completely performed in the segmentation step, multi-nucleic regions were considered as well segmented results. A region was considered ill segmented if: a) it was not detected or b) its geometry was not correctly identified. Table 1 presents the obtained detection accuracies on both datasets, when compared with the ground truth.

Table 1. Segmentation Accuracy Ratings

	Macrophages	*Parasites*
Segmentation total (DS_A)	3916	5257
Ground truth (DS_A)	4025	5572
Accuracy percentage (DS_A)	97.29	94.35
Segmentation total (DS_B)	4813	1832
Ground truth (DS_B)	5034	1981
Accuracy percentage (DS_B)	95.60	92.50

5.2   Classification Results

As dataset B possesses almost no testing data for either classifier due to its low clustering index, both classifiers were tested with data extracted from randomly selected regions in each of dataset A's images. Care was taken so that the data was divided, as equally as possible, between each class. The rule-based classifier's results are detailed bellow, in Table 2. No distinction was made between cellular or parasitic regions.

Table 2. Rule-Based Classifier Accuracy Ratings

Class	*0 Error Margin (correct)*	*±1 Error Margin*	*±2 Error Margin*	*±3 Error Margin*
0 (noise)	0.94	0.06	0.00	0.00
1	0.85	0.12	0.03	0.00
2	0.84	0.13	0.02	0.01
3	0.83	0.11	0.04	0.02
4	0.86	0.09	0.03	0.02
5	0.93	0.06	0.01	0.00
6	0.93	0.05	0.01	0.01
7	0.94	0.04	0.02	0.00
8	0.95	0.03	0.01	0.01
9	0.96	0.03	0.01	0.00

It is clear that as the number of nuclei increases, the classifier's error margins quickly decrease. The reason behind this is that, as the number of nuclei present in a region increases, the more the area value (normalized to the number of nuclei) approximates the distribution's mean value.

As previously mentioned, the log-likelihood ratios and their first and second order derivates were computed from mixture modeling 150 random clustered regions. These were then used to build several classifiers using Weka [15]. From these, Table 3 highlights the ones with the highest accuracy. All methods were validated through 2 and 10-fold cross validation (CV), as well as a 66% sequential split (SS). Ultimately, the SVM classifier was chosen due to its superior accuracy results. As with the RBC, care was taken so that the data was divided uniformly between classes.

Table 3. Machine-Learning Classifier Accuracy Ratings

Validation Type	C4.5 [19]	Best-First Search [19]	FFNN [19]	SVM [19]
Sequential Split	79.4%	82.4%	82.4%	**85.3%**
10-fold Cross Validation	79.4%	67.6%	70.6%	76.5%
2-fold Cross Validation	69.2%	69.2%	61.5%	69.2%

5.3   Execution Times

In order to assess the method's real-time capabilities, the running times for each step were computed in both datasets. Table 4 compares the mean running times of the proposed method, measured in seconds, with the mean annotation time for each dataset across subjects. The mean number of regions for dataset A was ~203.5 and ~607.3 for dataset B. The obtained results show that, even with a computationally taxing approach, the method is capable of analyzing images considerably faster than human experts. Thus, real-time annotation proves a feasible reality, albeit with room for small code improvements, such as parallelization techniques.

Table 4. Mean runtimes and standard deviation for the proposed methods various stages (in seconds). The tests were run on a laptop with an Intel Dual Core T2410 2,0 GHz CPU, 4 GB RAM, running Windows 7 32-bits. No graphical acceleration was used.

	Segmentation	Classification	Declustering	Total	Annotation
Dataset A	0.205 s	1.156 s	0.197 s	276.3 s	436.4 s
Dataset B	0.287 s	1.272 s	0.316 s	923.7 s	1395.7 s

6   Conclusions

This work has suggested a robust, automatic analysis methodology for cell and parasite detection in fluorescence microscopy imaging. The proposed method has shown itself robust to poor lighting conditions and high cluttering indexes, falling well within the error margins of expert biomedical researcher annotations.

The obtained results demonstrate this method is capable of performing the image analysis task adequately and in less time than a human expert. Being a computer program, the method also boasts from being immune to traditional human errors related to distraction, fatigue or subjectivity. Since these human errors are the major source of ambiguity in the manual annotation process, we consider our alternative to be a more suitable choice. This claim is supported with the fact that the method has a fixed error margin, meaning, the error does not randomly vary through time, as human errors do. Thus two or more drug trials can be safely compared as to their effectiveness, whereas if considering human error, the comparison requires validation through multiple confirmations.

Future work should focus on the methods used in the segmentation step, possibly employing mean shift or normalized cuts techniques, as well as increasing the training datasets of both classifiers. The built processing pipeline was made to be modular and applicable to other image types, thereby easily expandable to solve similar problems. This work has been successfully integrated with a pre-existing image annotation framework and is currently being used in the INEB/IBMC laboratories in Portugal.

References

[1] Ryan, K.J., Ray, C.G.: Sherris Medical Microbiology, pp. 749–754. McGraw-Hill (2004)

[2] Myler, P., Fasel, N.: Leishmania: After The Genome. Caister Academic Press (2008)

[3] Jeronimo, S.M.B., DeQueiroz-Sousa, A., Pearson, R.D.: Leishmaniasis. In: Goldman, L., Deds, A. (eds.) Cecil Medicine, 23th edn. ch. 369, Saunders Elsevier, Philadelphia (2007)

[4] Lichtman, J.W., Conchello, J.A.: Fluorescence Microscopy. Nature Publishing Group (2005)

[5] Spring, K.R.: MicroscopyU: Introduction to Fluorescence Microscopy (2010)

[6] Liao, Q., Deng, Y.: An Accurate Segmentation Method For White Blood Cell Images. In: Proceedings IEEE International Symposium on Biomedical Imaging, pp. 245–248 (2002)

[7] Ficarra, E., Cataldo, S.D., Acquaviva, A., Macii, E.: Automated Segmentation of Cells With IHC Membrane Staining. IEEE Transactions on Biomedical Engineering 58(5) (2011)

[8] Jiang, K., Liao, Q., Dai, S.: A Novel White Blood Cell Segmentation Scheme Using Scale-Space Filtering And Watershed Clustering. In: Proceedings of ICMLC (2003)

[9] Park, J., Keller, J.M.: Fuzzy Patch Label Relaxation in Bone Marrow Cell Segmentation. In: International Conference on Computational Cybernetics and Simulation, pp. 1133–1138 (1997)

[10] Begelman, G., Gur, E., Rivlin, E., Rudzsky, M., Zalevsky, Z.: Cell Nuclei Segmentation Using Fuzzy Logic Engine. In: Proceedings IEEE International Conference on Image Processing (2004)

[11] Yu, W., Lee, H.K., Hariharan, S., Bu, W., Ahmed, S.: Level Set Segmentation of Cellular Images Based on Topological Dependence. In: ISAVC (2008)

[12] Yan, P., Zhou, X., Shah, M., Wong, S.T.C.: Automatic Segmentation of High-Throughput RNAi Fluorescent Cellular Images. IEEE Transaction On Information Technology In Biomedicine 12(1) (2008)

[13] Otsu, N.: A threshold selection method from gray-level histograms. IEEE Trans. Sys., Man., Cyber. 9(1), 62–66 (1979)

[14] Freeman, H.: On the encoding of arbitrary geometric configurations. IRE Transactions on Electronic Computers, 260–268 (1961)

[15] Neal, R.A., Croft, S.L.: An in-vitro system for determining the activity of compounds against the intracellular amastigote form of Leishmania donovani. Journal of Antimicrobial Chemotherapy 14(5), 463–475 (1984)

[16] Hall, M., Frank, E., Holmes, G., Pfahringer, B., Reutemann, P., Witten, I.H.: The WEKA Data Mining Software: An Update. SIGKDD Explorations 11(1) (2009)

[17] Reynolds, D.: Gaussian Mixture Models. MIT Lincoln Laboratory, MA 02140, USA

[18] Gonzales, Woods: Digital Image Processing, 3rd edn. (DIP/3e) (2008)

[19] Bishop, C.M.: Pattern Recognition and Machine Learning. Springer (2007) ISBN: 0387310738

Link Prediction in Complex Networks Based on Cluster Information

Jorge Carlos Valverde-Rebaza and Alneu de Andrade Lopes

Departamento de Ciências de Computação
Instituto de Ciências Matemáticas e de Computação
Universidade de São Paulo - Campus de São Carlos
Caixa Postal 668
13560-970 São Carlos, SP, Brazil
{jvalverr,alneu}@icmc.usp.br

Abstract. Cluster in graphs is densely connected group of vertices sparsely connected to other groups. Hence, for prediction of a future link between a pair of vertices, these vertices common neighbors may play different roles depending on if they belong or not to the same cluster. Based on that, we propose a new measure (WIC) for link prediction between a pair of vertices considering the sets of their intra-cluster or within-cluster (W) and between-cluster or inter-cluster (IC) common neighbors. Also, we propose a set of measures, referred to as W forms, using only the set given by the within-cluster common neighbors instead of using the set of all common neighbors as usually considered in the basic local similarity measures. Consequently, a previous clustering scheme must be applied on the graph. Using three different clustering algorithms, we compared WIC measure with ten basic local similarity measures and their counterpart W forms on ten real networks. Our analyses suggest that clustering information, no matter the clustering algorithm used, improves link prediction accuracy.

Keywords: Link Prediction, Complex Networks, Clustering.

1 Introduction

Many social, biological, and information systems can be naturally described as networks, where vertices represent entities (individuals or organizations) and links denote relations or interactions between vertices [18], [30]. Networks or graphs are a powerful representation that has been employed in different tasks of machine learning (ML) and data mining (DM). This growing interest in the use of graph can be justified by the expressiveness of this representation and its applications include: supervised learning [16], [4], [19]; unsupervised learning [6], [25], [24], [20]; and semi-supervised learning [5], [3], [12], to cite just a few.

An important scientific issue regarding network analysis that has attracted increasing attention in recent years is the link prediction. The link prediction problem aims to estimate the likelihood of the future existence of a link between two disconnected vertices in a network, based on the observed links [13].

L.N. Barros et al. (Eds.): SBIA 2012, LNAI 7589, pp. 92–101, 2012.

Many methods for link prediction based on similarity between vertices have been proposed since similar vertices likely share the same relations (links). When the similarity between vertices is based solely on network structure, it is called structural similarity. Structural similarity measures can be classified in different ways, such as the based on local or global information, refer to [18] for details.

Liben-Nowell and Kleinberg [13] and Zhou *et al.* [30] systematically compared a number of structural similarity measures on real networks. According to the authors, global measure can provide higher accuracy, but its computation is very time-consuming and usually infeasible for large-scale networks, while local measure is generally faster but with lower accuracy.

Common Neighbors (CN) [17] is one of the simplest similarity measures based on local information that leads to a good performance. In networks with large clustering coefficient, i.e., if there are connections between a vertex a and two vertices b and c, probably there is a link between b and c, CN provides accurate predictions compared to measures based on global information [30]. The basic assumption of CN is that two vertices are more likely to be connected if they have more common neighbors. Thus, each common neighbor gives equal contribution to the connection likelihood. However, sometimes different common neighbors may play different roles and by identifying them may lead to more accurate prediction than CN [15]. For instance, the common friends in a same social group of two people who do not know each other may contribute more to their possibly future friendship than their common friends from different social groups.

Furthermore, in recent experiments on synthetic and real-world networks, Feng and colleagues found that for a network with low clustering structure link prediction measures based on structural similarity perform poorly. Nonetheless, as the clustering structure of the network grows, the accuracy of these measures drastically improves [9]. Inspired by these results, here, we firstly apply a partitioning scheme to divide the network into communities and then we explicitly use the obtained clustering structure information in the link prediction.

Considering that, to the connection likelihood between a pair of vertices, their common neighbors may contribute in different ways depending on if they belong or not to the same cluster, here we propose a new measure (WIC) for predicting link between a pair of vertices using information from intra-cluster or within-cluster (W) and inter-cluster (IC) common neighbors of these vertices. Also, considering solely the subset of within-cluster common neighbors instead of the set of all common neighbors, we also propose other measures, called *W forms*, derived from the basic local measures. Consequently, a clustering scheme must be applied on the graph analyzed before computing these measures. Thus, using three different clustering algorithms: FastQ (FQ) [6], an algorithm based on edge clustering coefficient (ECC) [25], and WalkTrap (WT) [24]. We compare the WIC measure with ten local similarity measures and their corresponding W forms on ten real networks. We show that cluster information about vertices, no matter the clustering algorithm used, improves the accuracy of link prediction over local similarity measures.

The remainder of the paper is organized as follows. In Section 2 we present the WIC measure. In Section 3 we present ten different local similarity measures and their corresponding W forms. In Section 4 we present experimental results on ten real networks drawn from disparate fields. Finally, in Section 5 we present the conclusions and discuss future work.

2   A Link Prediction Measure Based on Cluster Information

Consider an undirected network $G(V, E)$, where V is the set of vertices and E is the set of links. Multiple links and self-connections are not allowed. Consider the universal set, denote by U, containing all $\frac{|V| \cdot (|V|-1)}{2}$ possible links between vertices in V, where $|V|$ denotes the number of elements in V. The link prediction task is to find out the missing links (future links) in the set $U - E$ (set of nonexistent links) [30].

Given a pair of disconnected vertices (x, y), our task is to determine a similarity measure[1] that uses cluster information from the common neighbors of this pair of vertices. Consider that each vertex $v_i \in V$ is associated with a cluster label C that represents the cluster, community or any group of vertices that share some common properties and/or play similar roles within the network.

In network G exist $M > 1$ cluster labels $C_\alpha, C_\beta, \ldots, C_M$. When a vertex $x \in V$ belongs to a cluster with label C, this vertex is represented as x^C. Consider that a vertex belongs to a unique cluster.

Considering $\Gamma(x)$ denote the set of neighbors of vertex x, we denote by $\Lambda_{x,y} = \Gamma(x) \cap \Gamma(y)$ the set of common neighbors of the pair of disconnected vertices (x, y). According to Bayesian theory [11], the posterior probability that the same cluster label, C_α, be assigned to this pair of vertices, given their common neighbors $\Lambda_{x,y}$, is defined by Eq. 1.

$$P(x^{C_\alpha}, y^{C_\alpha} \mid \Lambda_{x,y}) = \frac{P(\Lambda_{x,y} \mid x^{C_\alpha}, y^{C_\alpha}) P(x^{C_\alpha}, y^{C_\alpha})}{P(\Lambda_{x,y})} \tag{1}$$

Similarly, the posterior probability that different cluster labels, C_α and C_β, be assigned to the vertices (x, y), given their common neighbors $\Lambda_{x,y}$, is defined by Eq. 2.

$$P(x^{C_\alpha}, y^{C_\beta} \mid \Lambda_{x,y}) = \frac{P(\Lambda_{x,y} \mid x^{C_\alpha}, y^{C_\beta}) P(x^{C_\alpha}, y^{C_\beta})}{P(\Lambda_{x,y})} \tag{2}$$

Eqs. 1 and 2 can not tell us which nonexistent links are more likely to exist than others. Nevertheless, we can derive an score measure for pairs of disconnected vertices (x, y) as the ratio of Eq. 1 to 2, as stated in Eq 3.

$$s_{x,y} = \frac{P(\Lambda_{x,y} \mid x^{C_\alpha}, y^{C_\alpha}) P(x^{C_\alpha}, y^{C_\alpha})}{P(\Lambda_{x,y} \mid x^{C_\alpha}, y^{C_\beta}) P(x^{C_\alpha}, y^{C_\beta})} \tag{3}$$

---

[1] We do not distinguish *similarity measure* and *score*.

Consider that $\Lambda_{x,y} = \Lambda_{x,y}^W \cup \Lambda_{x,y}^{IC}$, where $\Lambda_{x,y}^W = \{z \in \Lambda_{x,y} \mid x^C, y^C, z^C\}$ is the set of within-cluster (W) common neighbors and the complement $\Lambda_{x,y}^{IC} = \Lambda_{x,y} \setminus \Lambda_{x,y}^W$ is the set of inter-cluster (IC) common neighbors (common neighbors belonging to C_α, i.e., the same cluster of x, or C_β, the same cluster of y, or C_γ, any other cluster). Clearly, $\Lambda_{x,y}^W \cap \Lambda_{x,y}^{IC} = \varnothing$.

Hence, to estimate the probability of the common neighbors $\Lambda_{x,y}$ given x^{C_α} and y^{C_α}, we must consider the number of common neighbors with the same cluster label C_α by the total of common neighbors, i.e., the more the number of common neighbors in a same cluster the more the likelyhood of x and y belong to this cluster, Eq. 4.

$$P(\Lambda_{x,y} \mid x^{C_\alpha}, y^{C_\alpha}) = \frac{|\Lambda_{x,y}^W|}{|\Lambda_{x,y}|} \tag{4}$$

Similarly, to estimate the probability of the common neighbors $\Lambda_{x,y}$ given x^{C_α} and y^{C_β}, here we consider the number of common neighbors that may be associated with the cluster labels C_α or C_β or with another cluster label C_γ by the total of common neighbors, as stated in Eq. 5.

$$P(\Lambda_{x,y} \mid x^{C_\alpha}, y^{C_\beta}) = \frac{|\Lambda_{x,y}^{IC}|}{|\Lambda_{x,y}|} \tag{5}$$

Substituting Eqs. 4 and 5, the likelihood score of the pair of vertices x and y is

$$s_{x,y} = \frac{|\Lambda_{x,y}^W|}{|\Lambda_{x,y}^{IC}|} \times \frac{P(x^{C_\alpha}, y^{C_\alpha})}{P(x^{C_\alpha}, y^{C_\beta})} \tag{6}$$

The $\frac{P(x^{C_\alpha}, y^{C_\alpha})}{P(x^{C_\alpha}, y^{C_\beta})}$ ratio can be neglected since either this fraction value is 1 (when $\alpha = \beta$) leading $s_{x,y}$ to $\frac{|\Lambda_{x,y}^W|}{|\Lambda_{x,y}^{IC}|}$ or the score $s_{x,y}$ is 0 (when $\alpha \neq \beta$, in this case $\Lambda_{x,y}^W = \varnothing$).

Moreover, to prevent division by zero in the case when $\Lambda_{x,y}^W = \Lambda_{x,y}$ leading to $\Lambda_{x,y}^{IC} = \varnothing$ we add a small value constant $\delta \approx 0$ in denominator. The final score measure is computed by Eq. 7. We notice that $\delta \approx 0$ increases the score when $\Lambda_{x,y}^{IC} = \varnothing$, however it does not modify the evaluation based on the AUC measure, but it may modify the precision.

$$s_{x,y}^{WIC} = \frac{|\Lambda_{x,y}^W|}{|\Lambda_{x,y}^{IC}| + \delta} \tag{7}$$

3   Local Similarity Measures and Their W Forms

Different measures based on local information exist such as Common Neighbors (CN), Salton (Sal), Jaccard (Jac), Sørensen (Sor), Hub Promoted Index (HPI), Hub Depressed Index (HDI), Leicht-Holme-Newman index (LHN), Adamic-Adar

(AA), Resource Allocation (RA) and Preferential Attachment (PA), refer to [18] for details. All but PA measures use the set of common neighbors.

The simple counting of the number of common neighbors indicates that each common neighbor gives the same contribution to the connection likelihood. However, as already commented, different common neighbors may give different contributions to the connection probability [15].

Considering that within-cluster common neighbors may contribute more to the connection likelihood than inter-cluster common neighbors, we consider the subset of W common neighbors Λ_{xy}^{W} instead of using the set of all common neighbors Λ_{xy}, obtaining new measures referred to as W forms from the basic measures based on local information. Considering $k(x)$ is the degree of vertex x, Table 1 shows the local similarity measures and their corresponding W forms.

Table 1. Local similarity measures and their corresponding W forms

Local measure	W form
$s_{x,y}^{CN} = \|\Lambda_{xy}\|$	$s_{x,y}^{CN-W} = \|\Lambda_{xy}^{W}\|$
$s_{x,y}^{Sal} = \dfrac{\|\Lambda_{xy}\|}{\sqrt{k(x) \times k(y)}}$	$s_{x,y}^{Sal-W} = \dfrac{\|\Lambda_{xy}^{W}\|}{\sqrt{k(x) \times k(y)}}$
$s_{x,y}^{Jac} = \dfrac{\|\Lambda_{xy}\|}{\|\Gamma(x) \cup \Gamma(y)\|}$	$s_{x,y}^{Jac-W} = \dfrac{\|\Lambda_{xy}^{W}\|}{\|\Gamma(x) \cup \Gamma(y)\|}$
$s_{x,y}^{Sor} = \dfrac{2\|\Lambda_{xy}\|}{k(x)+k(y)}$	$s_{x,y}^{Sor-W} = \dfrac{2\|\Lambda_{xy}^{W}\|}{k(x)+k(y)}$
$s_{x,y}^{HPI} = \dfrac{\|\Lambda_{xy}\|}{min\{k(x),k(y)\}}$	$s_{x,y}^{HPI-W} = \dfrac{\|\Lambda_{xy}^{W}\|}{min\{k(x),k(y)\}}$
$s_{x,y}^{HDI} = \dfrac{\|\Lambda_{xy}\|}{max\{k(x),k(y)\}}$	$s_{x,y}^{HDI-W} = \dfrac{\|\Lambda_{xy}^{W}\|}{max\{k(x),k(y)\}}$
$s_{x,y}^{LHN} = \dfrac{\|\Lambda_{xy}\|}{k(x) \times k(y)}$	$s_{x,y}^{LHN-W} = \dfrac{\|\Lambda_{xy}^{W}\|}{k(x) \times k(y)}$
$s_{x,y}^{AA} = \sum_{z \in \Lambda_{xy}} \dfrac{1}{\log k(z)}$	$s_{x,y}^{AA-W} = \sum_{z \in \Lambda_{xy}^{W}} \dfrac{1}{\log k(z)}$
$s_{x,y}^{RA} = \sum_{z \in \Lambda_{xy}} \dfrac{1}{k(z)}$	$s_{x,y}^{RA-W} = \sum_{z \in \Lambda_{xy}^{W}} \dfrac{1}{k(z)}$
$s_{x,y}^{PA} = k(x) \times k(y)$	-

4   Experiments

We consider a scenario where new links of ten real networks from different fields must be predicted. In each one of these networks three different clustering algorithms are used, the FastQ (FQ) [6], an algorithm based on edge clustering coefficient, referred to as ECC [25] and WalkTrap (WT) [24], to assign a cluster label to each vertex. Finally, we compare the performance of our proposal and the W forms to the ones where similarity measures are based only on local information.

4.1   Datasets

The networks considered in our experiments are (i) Airline (AL) [2] (a network of US air transportation system), (ii) Football (FB) [10] (a network of American football games between Division IA colleges), (iii) Industry (IDT) [8] (a network of companies linked via cooccurrence in 35318 PR Newswire press), (iv) Karate (KT) [29] (a social network of friendships between members of a karate club), (v) Imdb (MN) [21](this network contains movies linked if they share a producer), (vi) NetScience (NS) [23] (a network of coautorship between scientists), (vii) Political Blogs (PB) [1] (a network of the US political blogs. The original links are directed, here we treat them as undirected links), (viii) Yeast (PPI) [27] (a protein-protein interaction network), (ix) Power (PW) [28] (an electrical power grid of western US), (x) Router (RT) [26] (a router-level topology of the Internet). The basic topological features [22] of these ten networks are summarized in Table 2.

4.2 Experimental Results

In this section, we present and discuss the results of using the WIC measure (using $\delta = 0.001$), the local similarity measures and their W forms. Each of the link prediction measures was implemented in C++ based on LPmade platform [14]. We use the experimental setup presented in [30], [15] and [18]. To test the prediction accuracy, the set of observed links, E, is randomly divided into two parts: the training set, containing 90% of the links, and the probe set, containing the remaining 10% of links. Table 3 summarizes the prediction accuracy results, measured by AUC on the ten networks whose vertices have a cluster label assigned by one of three different algorithms (FQ, ECC and WT).

In general, WIC measure has better performance on six of ten networks. We notice this performance is obtained on networks with large clustering coefficient and large degree of heterogeneity – defined as $\mathbf{H} = \langle k^2 \rangle / \langle k \rangle^2$, where $\langle k \rangle$ denotes the network average degree – such as AL, PPI, IDT and RT, where measures using the set of all common neighbors have few remarkable differences. In addition, in different networks, several W forms outperform, with significant difference, their corresponding basic forms. Note that, the performance of our propo- sals are observed independently of the clustering algorithm used. PA has the worst overall performance.

To show how the cluster information, no matter the clustering algorithm used, improves the link prediction accuracy, we present two different analyses from the results of Table 3. In the first, we analyze the difference between WIC measure and five local similarity measures. In the second, we analyze the statistical distributions of performance of all measures to emphasize the differences between the W forms and their corresponding basic forms.

The measures for link prediction mostly used in the literature are CN, Jac and AA. PA is interesting for requiring minimal information and RA due its similarity to AA. To analyze the difference between WIC measure and these five measures, we use a post-hoc test [7] from the results of the Table 3 and whose analysis is shown in three diagrams in Fig. 1. We show the critical difference (CD) on the top of each diagram. The axis in each diagram is the axis on which we plot the average ranks of measures. In the axis the lowest (best) ranks are in the left side. The measures analyzed have no significant difference, so they are connected by

Table 2. The basic topological features of ten experimental networks. Where $|V|$ and $|E|$ are the number of vertices and links. N_{C-FQ}, N_{C-ECC} and N_{C-WT} are the number of clusters found by the algorithms FQ, ECC and WT, respectively. Each entry for N_{C-FQ}, N_{C-ECC} and N_{C-WT} represents the total number of clusters detected and the number of clusters with a single node, for example, the entry 141/136 for AL with N_{C-FQ} means that the network has 141 clusters, detected by FQ algorithm, with 136 of these clusters are formed by a single node. **C** and **r** are clustering and assortative coeficient, respectively. **H** denotes the heterogeneity degree.

Nets	$|V|$	$|E|$	N_{C-FQ}	N_{C-ECC}	N_{C-WT}	C	r	H
AL	332	2126	141/136	5/0	32/17	0.6252	-0.2079	3.4639
FB	115	613	51/45	5/0	10/0	0.4032	0.1624	1.0069
IDT	2189	11666	1008/807	396/94	502/308	0.3297	0.1842	3.4122
KT	34	78	16/13	6/2	3/0	0.5706	-0.4756	1.6933
MN	1441	20317	793/523	321/282	359/315	0.5843	0.3492	2.0982
NS	1461	2742	740/524	269/0	1098/1082	0.6937	0.4616	1.8486
PB	1224	16716	598/584	122/92	13/2	0.3203	-0.2211	2.9749
PPI	2617	11855	1091/979	344/79	358/242	0.2844	0.461	3.7284
PW	4941	6594	2023/1985	738/175	234/0	0.0801	0.0034	1.4504
RT	5022	6258	1856/1786	165/2	208/0	0.0116	-0.1384	5.5031

Table 3. The prediction accuracy measured by AUC on ten networks. At each vertex of each network is assigned a cluster label using three different clustering algorithms (FQ, ECC and WT). Each AUC value is obtained by averaging over 100 implementations with independently random divisions of training set and probe set. The entries corresponding to the highest accuracies among the evaluated measures are emphasized by black.

	WIC	CN	CN-W	Jac	Jac-W	Sal	Sal-W	Sor	Sor-W	HPI	HPI-W	HDI	HDI-W	LHN	LHN-W	AA	AA-W	RA	RA-W	PA
AL (FQ)	**0.761**	0.669	0.617	0.737	0.717	0.725	0.756	0.668	0.759	0.672	0.755	0.675	0.756	0.677	0.739	0.663	0.756	0.661	0.757	0.535
AL (ECC)	**0.739**	0.669	0.668	0.737	0.697	0.725	0.648	0.668	0.669	0.672	0.674	0.675	0.615	0.677	0.68	0.663	0.589	0.661	0.585	0.535
AL (WT)	**0.738**	0.669	0.612	0.737	0.634	0.725	0.705	0.668	0.691	0.672	0.599	0.675	0.689	0.677	0.67	0.663	0.676	0.661	0.674	0.535
FB (FQ)	**0.929**	0.57	0.658	0.583	0.674	0.59	0.696	0.578	0.742	0.602	0.737	0.584	0.707	0.592	0.755	0.624	0.927	0.602	0.928	0.567
FB (ECC)	**0.839**	0.57	0.821	0.583	0.67	0.59	0.817	0.578	0.827	0.602	0.819	0.584	0.836	0.592	0.838	0.624	0.622	0.602	0.658	0.567
FB (WT)	**0.756**	0.57	0.731	0.583	0.585	0.59	0.706	0.578	0.741	0.602	0.682	0.584	0.728	0.592	0.714	0.624	0.754	0.602	0.749	0.567
IDT (FQ)	**0.729**	0.55	0.586	0.562	0.69	0.562	0.682	0.591	0.692	0.543	0.689	0.563	0.694	0.581	0.708	0.544	0.66	0.545	0.664	0.543
IDT (ECC)	**0.711**	0.55	0.659	0.562	0.656	0.562	0.674	0.591	0.653	0.543	0.637	0.563	0.649	0.581	0.614	0.544	0.655	0.545	0.662	0.543
IDT (WT)	**0.901**	0.55	0.58	0.562	0.883	0.562	0.882	0.591	0.725	0.543	0.711	0.563	0.705	0.581	0.667	0.544	0.743	0.545	0.813	0.543
KT (FQ)	0.657	0.739	0.605	0.697	0.776	0.796	0.705	0.742	0.515	0.756	0.605	0.683	0.515	0.708	0.515	0.832	**0.929**	0.836	**0.929**	0.866
KT (ECC)	0.736	0.739	0.742	0.697	0.796	0.796	**0.913**	0.742	0.838	0.756	0.75	0.683	0.75	0.708	0.805	0.832	0.661	0.836	0.643	0.866
KT (WT)	0.789	0.739	0.867	0.697	0.617	0.796	0.76	0.742	0.608	0.756	**0.929**	0.683	0.586	0.708	0.622	0.832	0.869	0.836	0.863	0.866
MN (FQ)	**0.722**	0.557	0.719	0.573	0.56	0.56	0.721	0.56	0.721	0.549	0.721	0.557	0.701	0.552	0.708	0.565	0.633	0.575	0.607	0.568
MN (ECC)	**0.63**	0.557	0.572	0.573	0.522	0.56	0.56	0.56	0.562	0.549	0.538	0.557	0.573	0.552	0.555	0.565	0.616	0.575	0.623	0.568
MN (WT)	**0.675**	0.557	0.673	0.573	0.567	0.56	0.657	0.56	0.667	0.549	0.633	0.557	0.664	0.552	0.587	0.565	0.622	0.575	0.606	0.568
NS (FQ)	0.743	**0.994**	0.864	0.979	0.848	0.99	0.864	0.993	0.894	0.991	0.823	0.992	0.793	0.985	0.796	0.983	0.633	0.909	0.81	0.784
NS (ECC)	0.774	**0.994**	0.869	0.979	0.793	0.99	0.827	0.993	0.827	0.991	0.877	0.992	0.823	0.985	0.75	0.983	0.781	0.909	0.812	0.784
NS (WT)	0.766	**0.994**	0.765	0.979	0.765	0.99	0.738	0.993	0.745	0.991	0.704	0.992	0.727	0.985	0.723	0.983	0.731	0.909	0.74	0.784
PB (FQ)	0.521	0.552	0.587	0.599	0.599	0.556	0.616	0.562	0.595	0.546	0.574	0.577	0.615	0.567	0.603	0.556	**0.628**	0.549	0.607	0.54
PB (ECC)	0.594	0.552	0.594	0.599	0.613	0.556	**0.642**	0.562	**0.642**	0.546	0.618	0.577	0.62	0.567	0.635	0.556	0.557	0.549	0.551	0.54
PB (WT)	0.511	0.552	0.512	0.599	0.513	0.556	**0.605**	0.562	0.561	0.546	0.547	0.577	0.523	0.567	0.524	0.556	0.556	0.549	0.536	0.54
PPI (FQ)	**0.904**	0.779	0.651	0.785	0.779	0.771	0.683	0.77	0.759	0.763	0.656	0.783	0.781	0.733	0.708	0.783	0.901	0.782	0.902	0.82
PPI (ECC)	**0.905**	0.779	0.899	0.785	0.784	0.771	0.833	0.77	0.804	0.763	0.78	0.783	0.852	0.733	0.693	0.783	0.8	0.782	0.784	0.82
PPI (WT)	**0.911**	0.779	0.892	0.785	0.761	0.771	0.795	0.77	0.815	0.763	0.782	0.783	0.808	0.733	0.576	0.783	0.785	0.782	0.782	0.82
PW (FQ)	0.575	0.575	0.575	0.575	0.505	0.555	0.555	0.57	0.505	0.565	0.505	0.575	0.505	0.555	0.505	0.575	**0.64**	0.575	0.637	0.45
PW (ECC)	0.581	0.575	0.571	0.575	0.635	0.555	0.674	0.57	0.643	0.565	**0.682**	0.575	0.635	0.555	0.67	0.575	0.581	0.575	0.578	0.45
PW (WT)	0.578	0.575	0.577	0.575	0.649	0.555	**0.656**	0.57	0.642	0.565	0.599	0.575	0.644	0.555	0.647	0.575	0.61	0.575	0.611	0.45
RT (FQ)	**0.909**	0.716	0.697	0.533	0.521	0.502	0.908	0.543	0.902	0.502	0.889	0.589	0.817	0.505	0.505	0.743	0.876	0.716	0.876	0.702
RT (ECC)	**0.76**	0.716	0.759	0.533	0.512	0.502	0.505	0.543	0.553	0.502	0.502	0.589	0.557	0.505	0.505	0.743	0.759	0.716	0.752	0.702
RT (WT)	**0.767**	0.716	0.765	0.533	0.526	0.502	0.505	0.543	0.523	0.502	0.511	0.589	0.537	0.505	0.505	0.743	0.765	0.716	0.759	0.702

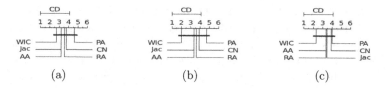

Fig. 1. Comparison of WIC measure with five measures (CN, Jac, AA, PA, RA) with a post-hoc test for results from Table 3. In (a), our proposal uses the information of the cluster labels assigned by the FQ algorithm in the ten networks analyzed. In (b), is used the ECC algorithm, and, in (c), is used the WT algorithm.

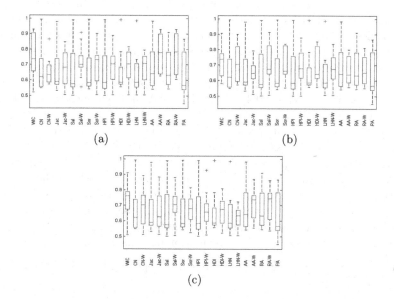

Fig. 2. Statistical distributions of performance of all measures analyzed. The line inside each box indicates the median of the prediction accuracies measured by AUC on ten networks. In (a), our proposals uses the information of the cluster labels assigned by FQ algorithm. In (b), is used the ECC algorithm, and, in (c) is used the WT algorithm.

a black line in each diagram. Although there is no significant difference between these measures, WIC measure has a better average ranking, which in turn is considerably larger than the average ranking of the next best measure (especially when using ECC and WT clustering algorithms). The average ranking of CN, Jac, AA and RA are closer to each other in the three diagrams. The PA has the worst overall average ranking.

Figure 2 shows the statistical distributions of performance of all measures analyzed. There is no single clear winner among the measures, however we can observe the following. First, all medians of accuracies obtained using W forms outperform their corresponding basic forms, no matter the clustering algorithm used, except for AA-W that is outperformed by a minimal margin by AA when using ECC algorithm. Second, while the median of WIC measure is always between 0.7 and 0.8, the medians of the W forms are always between 0.6 and 0.8, and the medians of the basic forms are always between 0.5 and 0.7. Furthermore, in most cases the difference between median of an W form and its basic form is considerable.

5   Conclusion

We proposed a new measure for link prediction in complex networks, called WIC measure. The WIC measure scores the likelihood of a link between a pair of vertices taking into account the clusters from which common neighbors of these vertices originated. This is, the measure uses the information denoted by high concentration of links within particular groups of vertices as well as by low concentration of links between these groups. Additionally, considering the subset of within-cluster common neighbors, we propose modifications to CN, Salton, Jaccard, Sørensen, HPI, HDI, LHN, AA and RA measures, obtaining their corresponding W forms.

Empirical analysis of our proposals compared with ten local similarity measures on ten real networks from different fields shows that there is no single clear winner but our proposals achieve better accuracies. Thus, the experiments carried out suggest that clustering information, independently of the clustering algorithm used, improves the link prediction accuracy. However the cost in the partitioning process must be considered. Finally, as social network analysis has become a hot topic in the last years, we intend to investigate how this proposal may be adapted to use semantics information of social networks.

Acknowledgements. This work is partially supported by CNPq agency.

References

1. Ackland, R.: Mapping the US political blogosphere: Are conservative bloggers more prominent? Presentation to BlogTalk, Downunder, Sydney (2005)
2. Batageli, V., Mrvar, A.: Pajek datasets (2006), http://vlado.fmf.uni-lj.si/pub/networks/data/mix/usair97.net
3. Bertini, J., Lopes, A., Zhao, L.: Partially labeled data stream classification with the semi-supervised K-associated graph. Journal of the Brazilian Computer Society, 1–12 (2012)
4. Bertini, J., Zhao, L., Motta, R., Lopes, A.: A nonparametric classification method based on k-associated graphs. Information Sciences 181(24), 5435–5456 (2011)
5. Blum, A., Chawla, S.: Learning from labeled and unlabeled data using graph mincuts. In: ICML, pp. 19–26 (2001)
6. Clauset, A., Newman, M.E.J., Moore, C.: Finding community structure in very large networks. Phys. Rev. E 70, 066111 (2004)
7. Demsar, J.: Statistical comparisons of classifiers over multiple data sets. JMLR 7, 1–30 (2006)

8. Fawcett, T., Provost, F.: Activity monitoring: Noticing interesting changes in behavior. In: Proc. of the Fifth ACM SIGKDD Int. Conf. on Knowledge Discovery and Data Mining, pp. 53–62 (1999)
9. Feng, X., Zhao, J.C., Xu, K.: Link prediction in complex networks: a clustering perspective. Eur. Phys. J. B 85(1-3) (2012)
10. Girvan, M., Newman, M.E.J.: Community structure in social and biological networks. PNAS 99(12), 7821–7826 (2002)
11. Hastie, T., Tibshirani, R., Friedman, J.: The elements of statistical learning: data mining, inference and prediction, 2nd edn. Springer (2009)
12. Laguna, V., Lopes, A.: Combining local and global knn with cotraining. In: ECAI 2010 - 19th European Conference on Artificial Intelligence, vol. 215, pp. 815–820. IOS Press, Netherlands (2010)
13. Liben-Nowell, D., Kleinberg, J.: The link-prediction problem for social networks. JASIST 58(7), 1019–1031 (2007)
14. Lichtenwalter, R.N., Chawla, N.V.: Lpmade: Link prediction made easy. JMLR 12, 2489–2492 (2011)
15. Liu, Z., Zhang, Q.-M., Lü, L., Zhou, T.: Link prediction in complex networks: A local naive bayes model. EPL 96(48007) (2011)
16. Lopes, A.A., Bertini Jr., J.R., Motta, R., Zhao, L.: Classification Based on the Optimal K-Associated Network. In: Zhou, J. (ed.) Complex 2009. LNICST, vol. 4, pp. 1167–1177. Springer, Heidelberg (2009)
17. Lorrain, F., White, H.C.: Structural equivalence of individuals in social networks. Journal of Mathematical Sociology 1, 49–80 (1971)
18. Lü, L., Zhou, T.: Link prediction in complex networks: A survey. Physica A: Statistical Mechanics and its Applications 390(6), 1150–1170 (2011)
19. Lu, Q., Getoor, L.: Link-based classification. In: ICML, pp. 496–503 (2003)
20. Motta, R., de Andrade Lopes, A., de Oliveira, M.C.F.: Centrality Measures from Complex Networks in Active Learning. In: Gama, J., Costa, V.S., Jorge, A.M., Brazdil, P.B. (eds.) DS 2009. LNCS, vol. 5808, pp. 184–196. Springer, Heidelberg (2009)
21. Neville, J., Jensen, D., Friedland, L., Hay, M.: Learning relational probability trees. In: KDD, pp. 625–630 (2003)
22. Newman, M.E.J.: The structure and function of complex networks. SIAM Review (45), 167–256 (2003)
23. Newman, M.E.J.: Finding community structure in networks using the eigenvectors of matrices. Phys. Rev. E 74, 036104 (2006)
24. Pons, P., Latapy, M.: Computing communities in large networks using random walks. J. Graph Algorithms Appl. 10(2), 191–218 (2006)
25. Radicchi, F., Castellano, C., Cecconi, F., Loreto, V., Parisi, D.: Defining and identifying communities in networks. PNAS 101(9), 2658 (2004)
26. Spring, N., Mahajan, R., Wetherall, D., Anderson, T.: Measuring ISP topologies with rocketfuel. IEEE/ACM Transactions on Networking 12(1), 2–16 (2004)
27. von Mering, C., Krause, R., Snel, B., Cornell, M., Oliver, S.G., Fields, S., Bork, P.: Comparative assessment of large-scale data sets of protein-protein interactions. Nature 417(6887), 399–403 (2002)
28. Watts, D.J., Strogatz, S.H.: Collective dynamics of small-world networks. Nature 393(6684), 440–442 (1998)
29. Zachary, W.W.: An information flow model for conflict and fission in small groups. Journal of Anthropological Research 33(4), 452–473 (1977)
30. Zhou, T., Lü, L., Zhang, Y.-C.: Predicting missing links via local information. Eur. Phys. J. B 71, 623 (2009)

A Parallel Approach
to Clustering with Ant Colony Optimization

Guilherme N. Ramos

Dept. of Computer Science, University of Brasília
gnramos@unb.br

Abstract. Recent innovations have enabled ever increasing amounts of data to be collected and stored, leading to the problem of extracting knowledge from it. Clustering techniques help organizing and understanding such data, and parallelization of such may reduce the cost of achieving this goal or improve on the result. This works presents the parallel implementation of the HACO clustering method, analyzing process of parallelization and its results with different topologies and communication strategies.

Keywords: clustering, ant colony optimization, hyperbox, parallel computing.

1 Introduction

Recent innovations have enabled the collection and storage of increasingly large volumes of data. The grouping of this data into clusters according to its characteristics, effectively dividing it into classes, leads to a number of advantages that can be exploited, such as:

- reducing the cost of collecting and sorting data;
- better performance in the classification of new data;
- discovery of patterns and relationships between attributes, or of attributes that are more relevant for classification.

Therefore, clustering may improve human understanding of the information and, thus, lead to the growth of the accumulated knowledge. Clustering methods, each with its own characteristics and applications, require significant effort to achieve reasonable results. One of the possible ways to attempt to reduce this cost is to use parallel implementations of the algorithms.

A method that could benefit from this approach is the *Hyperbox clustering with Ant Colony Optimization* (HACO), which uses hyperboxes and an inherently parallel meta-heuristic to achieve its goal [1]. This article aims to implement an easy to use system for intuitive clustering of data, improving and extending HACO, in order to exploit the method's optimization and parallelization features.

L.N. Barros et al. (Eds.): SBIA 2012, LNAI 7589, pp. 102–111, 2012.

2 Background on HACO

The *Hyperbox clustering with Ant Colony Optimization* method uses the ant colony meta-heuristic to near optimally place hyperboxes in the search space, clustering the data [1], considering the inherent topological information of the data.

A hyperbox defines a region in a space n-dimensional [2, 3], and is described by two vectors, usually their end points. Assuming an n-dimensional space of real numbers ($\Re^n$) and a hyperbox $H = (a, b)$, where $a \leq b$, then a point y is in H if:

$$H \subset \Re^n, y = \{y_1, y_2, \cdots, y_j, \cdots, y_n\}, y \in H \Rightarrow a_j \leq y_j \leq b_j, (a, b) \in \Re^n. \quad (1)$$

Where y_j is the j-th coordinate of y.

Hyperboxes represent an extremely simple concept that provides a direct interpretation for classifiers [1]. For example, the rule " if $a_j \leq y_j \leq b_j$, $\forall j \in [1, n]$ then the point y belongs to the class defined by hyperbox H", can be generated without any further calculations.

The ant colony optimization is a multi-agent system without centralized control, inspired by the behavior of ants in nature [4]. They communicate by depositing pheromone on the path between their nest and the source of food, and the collective behavior converges to the shortest path [4–7].

The idea behind HACO is simple: use these two concepts to cluster data. Each solution is modeled as the positions of hyperboxes in space, and the ant colony attempts to optimize the placement of these in order to maximize the amount of data contained within them [1], as illustrated in Figure 1b. The best solution found may be used in a strightforward manner to define clusters and extract simple classification rules. Though overlapping may occur, the hyperboxes tend to not be concentrated around the center of mass of data because once a data point has been accounted for in a hyperbox, it will not be considered for any other.

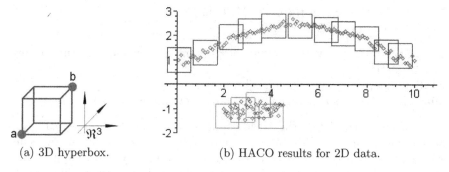

(a) 3D hyperbox.   (b) HACO results for 2D data.

Fig. 1. Hyperbox and hyperbox clustering examples

3   Evolving HACO

HACO was originally implemented in C language and it focused on solving the clustering problem, not in providing a flexible and reusable tool. Though the goal of this work is to have a parallel implementation of the method, other changes were made to improve its usability and flexibility.

3.1   Parallel Implementation

There are two reasons to seek a parallel alternative to solving a task [8], increasing the quality of the solutions within a fixed processing time slot or reducing the time required to reach a solution with adequate quality. The ant colony metaheuristic has inherently parallel characteristics [9], and - consequently - the same can be said of HACO.

The process of parallelization in the task of searching for a solution aims to define how to break down the problem in order to solve it as quickly as possible (or with minimal effort). It can be divided into two simple questions [10]:

- What can be done in parallel?
- When is the communication between parallel processes necessary?

The first question deals with how many pieces the problem can be broken into, there is no point in attempting a parallel application to a problem that cannot be divided; contrariwise, some problems can intuitively be split into smaller tasks for concurrent processing. The second question deals with the trade-off between the time gained by dividing the problem and the cost of communicating between different processes that are solving the pieces.

The Message Passing Interface [11] (MPI) defines a standard for data communication in parallel computing which proposes a single infrastructure for message exchange between multiple processors (or nodes) in a cluster [10, 12]. Though MPI is not the only approach to parallelization, it was chosen for this work due to its extensive online support and active community, and the fact that it was readily available within the Laboratório Nacional de Computação Científica (LNCC)'s infrastructure in the Open MPI[1] implementation.

The main features of MPI are:

Portability: processes with different programming languages can run on different platforms.
Usability: there are over 300 defined routines.
Availability: there is a huge variety of implementations available.

The operations that define the basic communication mechanism of MPI are *send* and *receive* [11]. This point-to-point communication, which involves an exchange of messages between two distinct tasks (and only two), happens between processes that use the exact same compiled code [10]. This differs from having the same sequential code running on different processors because: each

---

[1] http://www.open-mpi.org/

process can use its rank to determine a path in the flow of the algorithm, i.e. each process can use a particular piece of code; and the processes work together to execute a task.

To this end, the implementation of a program that uses MPI requires some important concepts [11, 13] to be known:

group: ordered set of processes, each with its own unique identifier (rank) in the $[0, N)$ interval for N processes.

context: abstract mechanism by which each process group differs in its communication with other system communications.

rank: identifier of the process relative to the group.

communicator: specifies the domain of communication of a set of processes (selects the group and the context of communication).

data type: describes the information being transferred.

request: identifies communication operations.

status: used to assess the situation of an operation.

Since the main purpose of a group and its communicators is to enable organization of tasks into classes of tasks, allowing collective communication operations over subsets of related tasks, and to provide secure communication, there are several possibilities for a parallel version of the ant colony algorithm [14].

Independent parallel colonies, each acting sequentially on a process (there is no communication cost between them), aim to increase the quality of the solutions within a fixed processing time slot. Communicating parallel ants, each with its own process, work together in search of solution, exchanging information with each other during the processing. The same idea can be applied with parallel colonies, and both these approaches aim to reduce the time required to reach a solution with appropriate quality. Finally, there can be parallel evaluation of solutions, generally the most expensive part of processing, which also aims to reduce the time required to reach a solution.

3.2   Object-Oriented Design

In order to make it easier to implement changes and maintain the code, HACO was re-implemented in C++, aiming to isolate the components and exploit the benefits of an Object-oriented (OO) design. In addition to the paradigm's features, which accommodate the problem's natural breakdown into a collection of data structures and operations on these structures [10], C++ generic programming was used to implement versions of the algorithms in a more flexible manner. Compared to the original version, not only does this result in a cleaner, more structured code, but it also makes it much simpler to apply the method to different data sets (broadening its range of applications).

The benefits of this could be seen in ant colony optimization's algorithm's *run* method, which is called in every iteration. In this case, each method is defined according to its use, but the algorithm's behavior remains constant. Thus, the same algorithm can be applied to different problems (such as the traveling salesman problem [4, 15] or traffic routing in communication networks [16]) or have different implementations to solve the same problem (such as data clustering in [1, 17]).

```
start_run();
while(!stop_run) {
    generate_solutions();
    store_best_solution_in_run();
    update_pheromone();
}
```

Fig. 2. Ant Colony Optimization algorithm

By using class hierarchy, the class MPIBase defines the basic characteristics of communication using MPI, and the class Ring, derived from MPIBase implements a ring topology for passing messages. This feature was also used to join the two distinct parts of the work proposed here: classes that deal with the problem itself (HACO) and classes that deal with communication (such as class Ring) are parents of a class that does both (HACO_Ring_DuringRun), as shown in Figure 3.

4   HACO – A Parallel Approach

4.1   Object Oriented Implementation

The initial step was to define the class Algorithm using generic programming in order to standardize the behavior for comparison of different algorithms. This template class enables any algorithm to be easily evaluated and compared to others. Basically, the class has the method *execute*, which calls the *run* method that must be implemented by sub-classes, computing the time cost for executions. Class Clusters defines characteristics for clustering algorithms, such as implementation and manipulation of a data structure which deals with the centroid of each group for analyzing the solution.

The logic behind the ant colony optimization, as illustrated in Figure 2, is implemented in the class ACO. This class provides an implementation of the meta-heuristic that is generic enough to solve any combinatorial problem that can be modeled. The classes Algorithm, Clusters, and ACO, when put together, describe the expected behavior of an algorithm that groups data using the ant colony optimization. Their use greatly simplifies the implementation of such a grouping class, with any clustering algorithm, and the associated tasks of performance tests, as was done in class HACO, which implements the logic for the *Hyperbox clustering with Ant Colony Optimization* method.

Such modularity was also taken advantage of in the parallel implementation of the algorithm. Starting from a base class MPIBase, which defines the characteristics for communication via MPI and the necessary interfaces, specialized sub-classes that only define the topology communication network can be created (e.g. classes `Star` and `Ring`). These are easily integrated with the class that handles the logic of the process (e.g. `HACO`), and the resulting class is a straightforward implementation of a parallel version of the method. This is exemplified in the `HACO_Ring_DuringRun` class, illustrated in Figure 3.

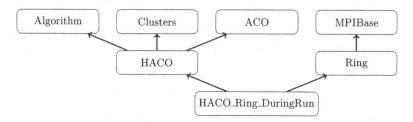

Fig. 3. `HACO_Ring_DuringRun` class diagram

The implementation of the `Algorithm`, `ACO`, and `MPIBase` classes follows the design pattern *facade* to provide an interface for the desired functionality [18] while the `Clusters` class defines methods and attributes to handle data clustering.

4.2 Parallelization of HACO

The most straightforward way to parallelize an implementation of the ant colony optimization algorithm such as HACO is to use a separate process for each agent and a master process to update the pheromone with the best solution found [19]. At every iteration the pheromone matrix is sent to each process, a procedure that can be costly. To reduce communication overhead, this update can be made after a predetermined number of iterations [20], an implementation that aims to reduce the time necessary to solve the problem.

Another way to implement a parallel version of HACO is to use each process as a distinct attempt to find the solution, and ultimately the master process selects the best solution among the available. The purpose of this parallel search for colonies is to decrease the time necessary for finding a solution.

5   Experimental Results

The cluster used for this work at LNCC consists of 72 machines (execution nodes), each with the following specification:

- Model Sun Blade
- 2 Intel Xeon E5440 Quad Core

- 16 GB PC2-5300 DDR2
- CentOS 5.3 operating system

To use it, the user accesses the master machine and submits the processes to the queue manager Sun Grid Engine. The cluster uses the environment modules package to dynamically modify the user's environment, configuring the shell. In the experiments, using the *openmpi* module, the specified parallel environment (PE) was mpi1, which defines a process per machine using MPI.

HACO was used to sort the data shown in Figure 1b, which consist of 150 computer-generated random data points which and can, intuitively, be separated into two distinct groups. This problem's purpose is simply to provide a simple benchmark for comparison of different implementations, it is small enough to not be costly, but complex enough to test the clustering algorithm. The method was applied with the following parameter values (see [1] for details of each):

Table 1. HACO parameters for the experiments

Parameter		Value
Iterations	I	100
Number of agents	R	10
Probability of exploration	P_E	0.98
Probability of local optimization	P_{LS}	0.1
Persistence of pheromone	ρ	0.9
Ratio of space to search	α	0.5
Length of the edge	D'	1
Number of solutions to upgrade	L	1

HACO is a method better suited for problems with NP complexity and unknown optimal answers. Thus, and since time (number of iterations) is already a stopping criterion, the goal is to improve the quality of the solution within the allotted period. Two different topologies were used for communication, the ring, in which each process communicates through MPI with the adjacent process, and the star, in which a master process collects the best solutions of all other processes, sorts them and propagates the best one back to them. HACO was executed 10 times in each of the five processes specified.

Furthermore, the communication during two distinct moments was tested for each topology. The first considered independent executions of the algorithm, i.e. the communication between was made after each independent process was finished to select the best solution found. In the second, communication happened within each process as it is being executed. These communication styles are shown in Table 2 with the letters IR (independent runs) and DR (during run), respectively. For communication during the run, a message is sent in every 25 iterations. These results were compared to the a single run of the HACO method without any parallel implementation. The fitness represents the number of data points contained withing hyperboxes.

Figures 4a and 5a show that the best solution for all processes is passed to each process, aiming to guide its pheromone matrix update according to the fittest case so far. Figures 4b and 5b show that the method does find reasonable solutions quickly and independently, but that the average fitness is less than the implementation that communicates during the runs.

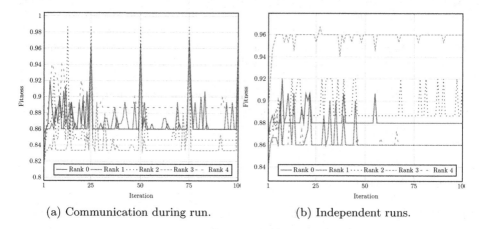

(a) Communication during run.   (b) Independent runs.

Fig. 4. Solution fitness for the Star topology

Comparing HACO in its original implementation to the parallel versions, it can be seen that although both reach acceptable results, on average the parallel versions provide better solutions. The trade-off, as expected, is that for this

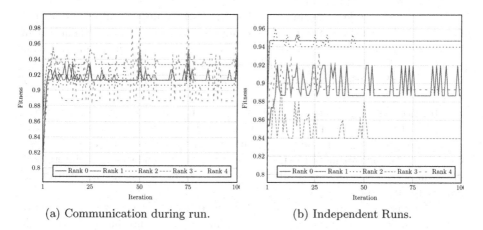

(a) Communication during run.   (b) Independent Runs.

Fig. 5. Solution fitness for the Ring topology

Table 2. Experimental results

Topology	Communication	Best Fitness	Avg. Fitness	Avg. Time (ms)
Ring	During Run	0.987	0.954	2440
	Independent Runs	0.987	0.943	1763
Star	During Run	0.987	0.967	2522
	Independent Runs	0.987	0.943	2405
Original HACO		0.987	0.933	1647

experimental setup, the cost of communication is high, as shown by the time required for execution.

The Ring topology require less messages being transmitted than the Star topology, resulting in a faster implementation with similar fitness results. Communicating solutions with the Star topology during runs, however, provided the highest average fitness. This makes sense since the ant colony optimization guides its search through the pheromone matrix, which is updated according to the best solutions found. The communication during the search process propagates the best solution found (by all processes), redirecting the pheromone updates in all processes.

6   Conclusions

The HACO method takes a different approach to clustering, aiming to structure the data and provide information on this for analysis for application in the ever increasing volume of data being collected nowadays. A parallel version is proposed to exploit the method's nature when searching for a solution.

In order to use parallel communication between HACO processes, the method was reimplemented in a much more modular and flexible OO manner. Exploiting the advantages of this paradigm, MPI classes were created and easily integrated to the existing method for a flexible, general-purpose clustering algorithm that can be run in parallel machines.

Results on a simple problem show that the parallel implementation does not impose much overhead on the serial version, so both can be considered equivalent when applied to a single process. On the other hand, certain communication patterns, such as star topology with messages being sent during the process, showed that the parallel implementation has potential to surpass the results of the original implementation.

Indications of the changes in the fitness of the solution can be seen in the evolution of the solution (Figures 4 and 5), however they do not seem to have caused enough changes in the pheromone matrix to redirect the search process. This issue is probably due to the update mechanism, which leads to an interesting question: should the shared solution have a stronger weight in the pheromone matrix update process?

HACO has been compared to standard clustering algorithms when applied to real case studies [1], with successful results. Since there are no changes in the logic for building a solution, the parallel version should provide results with similar quality, but this can only be verified with more experiments. To this end, and aiming to explore the method's flexibility and practical usability, applications to real data are being pursued. Implementations with different topologies and communicating strategies are also being examined.

Acknowledgments. This work was supported by Brazilian funding from CNPq and PCI/MCT. The author would like to thank Dr. Helio J.C. Barbosa and the staff at LNCC for their support during the research period.

References

1. Ramos, G.N., Hatakeyama, Y., Dong, F., Hirota, K.: Hyperbox clustering with Ant Colony Optimization (HACO) method and its application to medical risk profile recognition. Applied Soft Computing 9(2), 632–640 (2009)
2. Simpson, P.K.: Fuzzy Min-max Neural Networks – Part 1: Classification. IEEE Transactions on Neural Networks 3(5), 776–786 (1992)
3. Simpson, P.K.: Fuzzy Min-max Neural Networks – Part 2: Clustering. IEEE Transactions on Fuzzy Systems 1(1), 32–45 (1993)
4. Dorigo, M., Stützle, T.: Ant Colony Optimization. MIT Press (2004)
5. Deneubourg, J.L., Aron, S., Goss, S., Pasteels, J.M.: The Self-organizing Exploratory Pattern of the Argentine Ant. Journal of Insect Behavior 3(2), 159–168 (1990)
6. Dorigo, M., Maniezzo, V., Colorni, A.: Ant System: Optimization by a Colony of Cooperating Agents. IEEE Transactions on Systems, Man, and Cybernetics, Part B 26(1), 29–41 (1996)
7. Goss, S., Aron, S., Deneubourg, J.L., Pasteels, J.M.: Self-organized Shortcuts in the Argentine Ant. Naturwissenschaften 76(12), 579–581 (1989)
8. Manfrin, M., Birattari, M., Stützle, T., Dorigo, M.: Parallel Ant Colony Optimization for the Traveling Salesman Problem. In: Dorigo, M., Gambardella, L.M., Birattari, M., Martinoli, A., Poli, R., Stützle, T. (eds.) ANTS 2006. LNCS, vol. 4150, pp. 224–234. Springer, Heidelberg (2006)
9. Ellabib, I., Calamai, P., Basir, O.: Exchange strategies for multiple Ant Colony System. Information Sciences 177(5), 1248–1264 (2007)
10. Karniadakis, G.E., Kirby, R.M.: Parallel Scientific Computing in C++ and MPI. Cambridge University Press (2003)
11. MPI: A Message-Passing Interface Standard Version 2.2., http://www.mpi-forum.org/docs/mpi-2.2/mpi22-report.pdf (online; accessed August 2010)
12. Alba, E., Tomassini, M.: Parallelism and evolutionary algorithms. IEEE Transactions on Evolutionary Computation 6(5), 443–462 (2002)
13. Skjellum, A., Lu, Z., Bangalore, P.V., Doss, N.: Explicit Parallel Programming in C++ based on the Message-Passing Interface (MPI). Parallel Programming Using C++, 767–776 (1995)
14. Randall, M., Lewis, A.: A Parallel Implementation of Ant Colony Optimization. Journal of Parallel and Distributed Computing 62(9), 1421–1432 (2002)
15. Gambardella, L.M., Dorigo, M.: Ant-Q: A Reinforcement Learning Approach to the Traveling Salesman Problem. In: Proceedings of the Twelfth International Conference on Machine Learning, pp. 252–260 (1995)
16. Di Caro, G., Dorigo, M.: AntNet: Distributed Stigmergetic Control for Communications Networks. Journal of Artificial Intelligence Research 9(2), 317–365 (1998)
17. Shelokar, P.: An ant colony approach for clustering. Analytica Chimica Acta 509(2), 187–195 (2004)
18. Martin, R.C.: Agile software development: principles, patterns, and practices. Prentice Hall PTR Upper Saddle River, NJ (2003)
19. Bullnheimer, B., Kotsis, G., Strauss, C.: Parallelization strategies for the ant system. Report Series SFB "Adaptive Information Systems and Modelling in Economics and Management Science 8 (1997)
20. Antony, D., Piriyakumar, L., Levi, P.: A new approach to exploiting parallelism in ant colony optimization. In: Proceedings of 2002 International Symposium on Micromechatronics and Human Science, pp. 237–243 (2002)
21. Schildt, H.: C, The complete reference, 4th edn. Osborne/McGraw-Hill (2000)

On the Use of Consensus Clustering
for Incremental Learning of Topic Hierarchies

Ricardo M. Marcacini, Eduardo R. Hruschka, and Solange O. Rezende

Mathematical and Computer Sciences Institute - ICMC
University of São Paulo - USP
São Carlos, SP, Brazil
{rmm,erh,solange}@icmc.usp.br

Abstract. Incremental learning of topic hierarchies is very useful to organize and manage growing text collections, thereby summarizing the implicit knowledge from textual data. However, currently available methods have some limitations to perform the incremental learning phase. In particular, when the initial topic hierarchy is not suitable for modeling the data, new documents are inserted into inappropriate topics and this error gets propagated into future hierarchy updates, thus decreasing the quality of the knowledge extraction process. We introduce a method for obtaining more robust initial topic hierarchies by using consensus clustering. Experimental results on several text collections show that our method significantly reduces the degradation of the topic hierarchies during the incremental learning compared to a traditional method.

Keywords: topic hierarchies, consensus clustering, incremental learning.

1 Introduction

Topic hierarchies are efficient data models to organize and analyze the implicit knowledge from textual collections [12,15,1]. In these models, text documents are organized into topics and subtopics, providing an intuitive way for users to explore textual data at different levels of granularity [12,20]. Moreover, topic hierarchies allow the addition of new knowledge in a incremental manner, since new documents can be inserted into preexisting topics [13,15]. Due to both the rapid growth of online textual repositories and the need to process large volumes of texts, topic hierarchies have become an important tool in several applications, such as for digital libraries [17], webmining [12], and search engines [4].

The construction of topic hierarchies by using supervised methods usually requires an intense human effort, in particular to building the model [12,15,11]. Furthermore, those methods become impracticable for growing text collections. Thus, unsupervised methods for learning topic hierarchies from texts have received great attention in the literature [1,15,8]. In particular, hierarchical text clustering algorithms are potentially useful because they allow automatic organization of textual collections into clusters and subclusters based on similarity

L.N. Barros et al. (Eds.): SBIA 2012, LNAI 7589, pp. 112–121, 2012.
© Springer-Verlag Berlin Heidelberg 2012

measures between documents [16,20]. Each (sub)cluster contains documents that are conceptually similar in the sense that they deal with related topics.

An interesting approach for text clustering is known as Buckshot clustering [5,16]. Buckshot clustering works in two phases [6,16]: (1) a traditional hierarchical clustering algorithm is applied on a random sample from a textual collection — by doing so, an initial model is built; (2) the remaining documents are then assigned to the hierarchy built in phase (1). This is done based on the similarity measure between the incoming document and cluster centroids. While the first phase builds an initial model in a offline way, the second phase makes it possible to perform incremental model updates as new data is available. From this viewpoint, it can be observed that the quality of the initial model is a critical issue for incremental phase. In general, the initial model should preserve the cluster structure of the whole textual collection. Otherwise, new documents are inserted into wrong clusters and this error gets propagated into future hierarchy updates, thus decreasing the quality of the knowledge extraction process. This is a common problem with incremental learning [3] and exploring ways to reduce its effects is useful for various applications.

We introduce a method, called **Buckshot Consensus Clustering (BC²)**, for obtaining more robust initial topic hierarchies by using consensus clustering integrated with the Buckshot approach. Consensus clustering combines different clustering solutions from a same dataset into a single clustering solution with (hopefully) better quality [18,9]. For example, if a document is misplaced in some clustering solution, the same document is not necessarily misplaced in other clustering solutions, thereby consensus clustering can yield to better final solutions (data partitions). An experimental evaluation on several text collections shows that our method BC² significantly reduces the degradation of the topic hierarchies during the incremental learning compared to a traditional Buckshot approach. Furthermore, our BC² can even obtain topic hierarchies with quality similar to other non-incremental methods. This result indicates that the proposed BC² method is a promising alternative, particularly for reducing the effort for maintenance of topic hierarchies in real and dynamic environments.

The reminder of this paper is organized as follows. The next section presents related work on unsupervised learning of topic hierarchies. Section 3 describes the proposed method for incremental learning of topic hierarchies with consensus clustering. An experimental evaluation is carried out and the results are discussed in Section 4. Finally, Section 5 presents conclusions and future works.

2   Related Work

Several studies in the literature have investigated how to learn topic hierarchies by using text clustering. The Scatter/Gather approach [5] is one of the pioneering studies, where a cluster-based document browsing method allows users to interactively explore document collections. In a first step, a small number of topics is presented to the user (Scatter). Then, in a second step, the user selects the topic of interest that will be refined into subtopics (Gather). In this context, the Buckshot algorithm was proposed to deal with the fast clustering steps required by the Scatter/Gather. In another study, the Scatter/Gather (SG) steps was adapted and evaluated for learning topic hierarchies in an unsupervised way [11]. Over the years, variations of the SG have been proposed — *e.g.*, online and parallel versions for the Buckshot algorithm [10,1].

The extraction of topics from document clusters is performed by cluster labeling techniques. In general, studies based on the SG approach select the most frequent terms of each cluster to identify the topics. More advanced approaches have been proposed and include the use of probabilistic models based on term co-occurrences [8] and latent topics [19]. In addition, there are topic extraction approaches that explore feature selection techniques for hierarchical clustering [14]. The advantage of the feature selection techniques is that any hierarchical clustering algorithm can be used. In this context, an experimental evaluation performed by [20] indicated that the algorithms UPGMA and Bisecting k-means are the most recommended ones for hierarchical text clustering, and therefore are good choices for obtaining the initial clustering in the Buckshot algorithm.

Finally, the incremental learning phase is usually based on incremental clustering algorithms. The main objective is to insert new documents into existing clusters without having to reprocess previously clustered documents. Most existing incremental clustering algorithms use some variant of the Leader-Follower algorithm [16], where each new document is inserted into the closest cluster by calculating the similarity between the new document and the cluster centroids. Like any incremental learning method, a potential issue in incremental clustering is the stability-plasticity dilemma [3], i.e., some previously learned knowledge may have to be lost in order to learn new knowledge. A completely stable model will preserve the existing knowledge, but it will not be able to incorporate new knowledge. Similarly, completely plastic models have the ability to incorporate new knowledge, but without preserving the previously knowledge. Existing studies on incremental learning of topic hierarchies are limited in exploring the properties of an incremental clustering algorithm in such a way that a tradeoff between stability and plasticity can be obtained [13,3,15]. In this work, we explore an underexplored aspect, namely: how to improve the quality of the initial model by using consensus clustering (instead of using a single initial clustering solution), and its effects in the incremental learning phase. It is worth mentioning that consensus clustering has shown to be helpful for non-incremental methods [2,18,9]. This observation has motivated us to employ related approaches for obtaining better initial topic hierarchies, which in this particular setting is an aspect not addressed in the literature.

3   Buckshot Consensus Clustering (BC^2)

We now describe a method for unsupervised learning of topic hierarchies based on document sampling and consensus clustering. We adopt an approach similar to the ones used by classical clustering algorithms based on random sampling — like Buckshot [5,11,6] and CURE (*Clustering Using Representatives*) [7,16].

The proposed BC^2 method is divided into three main steps: document sampling, initial model building, and incremental model updating.

In the first step, **document sampling** is used to reduce the size of the data employed to build the initial model. Empirical and analytical studies in the literature report that random sampling can preserve the main cluster structures from the entire data set [7] (provided that the sample is large enough). However, the choice of the sample size is not trivial and depends on several factors, such as computational resources (*e.g.*, main memory), desired number of clusters, and the total number of documents in the textual collection.

In this work, we use the *Chernoff Bounds* for random sampling in clustering tasks, as introduced in [7], to set the sample size. Thus, given a collection of n documents, the size s of the document sample is calculated by using the Eq. (1):

$$s \geq fn + klog(\frac{1}{\delta}) + k\sqrt{log(\frac{1}{\delta})^2 + 2f(\frac{n}{k})log(\frac{1}{\delta})} \qquad (1)$$

where k is the desired number of clusters, f is a fraction of the number of documents to be sampled in each cluster ($0 \leq f \leq 1$), and δ is the confidence level of the sampling process ($0 \leq \delta \leq 1$). Therefore, the probability that the document sample of size s contains fewer that $f(\frac{n}{k})$ documents for each cluster is less than δ [7].

After selecting a random sample of documents, we proceed to the second step: **initial model building**. The potential advantage of this step, compared to traditional methods, is the use of consensus clustering to obtain a more robust initial model. Given a sample S_0 with s documents, C data partitions are obtained from S_0 by running various clustering algorithms or alternatively repeated runs of the same algorithm with different parameter values. Assuming that similar documents are in the same clusters in various of the (different) clustering solutions, then we compute the so-called co-association matrix [18]. The basic idea is to summarize the set of C data partitions by means of a matrix whose elements are $M(i,j) = \frac{a_{ij}}{|C|}$, where a_{ij} is the number of times that the documents i and j are in the same cluster, considering all the clusters from the C partitions. The co-association matrix $M_{|S_0| \times |S_0|}$ represents new relationships between documents obtained from the random sampling procedure. In order to finish this step, a hierarchical clustering H is obtained from the co-association matrix. This way, the various partitions are combined into consensus clustering. Then, one can use any technique for topic extraction from hierarchical clustering, H, thereby resulting in the initial model of a new topic hierarchy, T.

The third and last step of our BC^2 method is the **incremental model updating**, which allows incremental learning of topic hierarchies. Given an incremental source of documents and a topic hierarchy T (obtained in the previous step), we use the *Leader-Follower* algorithm to insert the remaining documents into T. Thus, for each new document, a top-down search is performed in the topic hierarchy to find the leaf topic that is the most similar to the new document. Then, the document is inserted into the selected topic, and the cluster center of this topic, as well as all of its ancestors, are updated with the information from the new document. This step continues until there are no more documents to be inserted, thereby obtaining an updated topic hierarchy.

4   Experimental Evaluation

We carried out an experimental evaluation to analyze the impact of the use of consensus clustering in the incremental learning setting of topic hierarchies. Our BC^2 method was compared with the traditional Buckshot approach, and also with a non-incremental method. The textual collections, as well as the algorithms for learning topic hierarchies from texts, are available in our software toolkit called *Torch - Topic Hierarchies*[1].

4.1   Document Collections

We used a total of 10 benchmark textual collections from various sources. Table 1 shows a summary of these textual collections. The smallest collections contain 2,301 documents, while the largest one contains 18,828 documents. All text documents were preprocessed by following the recommendations of [20]: (i) *stop-words* removal, such as pronouns, prepositions and articles; (ii) term *stemming* by Porter's algorithm, in which variations of a word are reduced to their radical; and (iii) removal of terms that occur in less than two documents.

Table 1. Summary of document collections used in the experimental evaluation

Dataset	Source	#Terms	#Documents	#Categories
20ng	Newsgroup messages	18,745	18,828	20
ACM	Computer science papers	3,462	3,498	40
Hitech	San Jose Mercury (TREC)	2,289	2,301	6
LATimes	LA Times (TREC)	6,141	6,279	6
NSF	National Science Foundation	10,160	10,521	16
Oshumed400	Heart diseases papers	6,974	9,200	23
PubMed-Cancer	Cancer research papers	15,559	15,588	12
Re8	Reuters-21578	7,555	7,674	8
Reviews	San Jose Mercury (TREC)	4,047	4,069	5
WebACE	Yahoo! Directory	3,857	3,900	21

In all textual collections, the documents are organized into categories of reference. These can be used as ground truth partitions. Thus, it is possible to evaluate the performance of the topic hierarchies by using objective measures, like the one addressed in the next section.

4.2   Evaluation Criteria

We used the F_{SCORE} measure [11,20] to quantitatively assess the quality of the obtained topic hierarchies. It is essentially an information retrieval measure that computes how much the topic hierarchy can recover the category information associated with each document. For this purpose, consider the following:

---

[1] Torch - Topic Hierarchies: http://sites.labic.icmc.usp.br/torch/

- T is a topic hierarchy obtained by an unsupervised learning method;
- Q_i is a topic belonging to T, where Q_i contains a set of documents; and
- L_r is a category of reference and its respective set of documents.

The F measure of a category L_r is calculated by choosing the maximum value obtained in some topic $Q_i \in T$, according to Eq. (2). In this case, $F(L_r, Q_i)$ is a mean between precision $P(L_r, Q_i) = \frac{|L_r \cap Q_i|}{|Q_i|}$ and recall $R(L_r, Q_i) = \frac{|L_r \cap Q_i|}{|L_r|}$. The F_{SCORE} of a topic hierarchy with n documents and c categories is the sum of the F values of the categories weighted by the size of each category (Eq. 3).

$$F(L_r) = \max_{Q_i \in T} F(L_r, Q_i) \quad (2) \qquad F_{SCORE} = \sum_{r=1}^{c} \frac{|L_r|}{n} F(L_r) \quad (3)$$

Thus, if the topic hierarchy perfectly recovers the information category of the documents, then the F_{SCORE} measure is equal to 1. In general, the higher the F_{SCORE} values, the better the topic hierarchy.

4.3   Experimental Setup

In the experimental setup, each document collection was divided into two sets: (1) S_0 with a sample of documents for learning the initial model, and (2) X_{inc} with the set of documents used for incremental updating of the model.

The X_{inc} sets were divided into 10 subsets, where $X_{inc} = S_1 \cup S_2 \cup ... \cup S_{10}$ and $S_1 \cap S_2 \cap ... \cap S_{10} = \emptyset$. Thus, we can evaluate the model degradation in 10 different stages of the incremental learning process. We used four methods to learn topic hierarchies:

1. **UPGMA+Incremental:** it represents the traditional Buckshot approach in which the initial model is obtained by the classic clustering algorithm UPGMA [6,16]. The incremental model updating step is based on the Leader-Follower algorithm [16] — using the cosine similarity measure;
2. **Consensus+Incremental (our BC^2 method):** it uses consensus clustering for learning the initial model, as discussed in Section 3;
3. **Offline Clustering (topline):** used to learn a topic hierarchy in a non-incremental way, with access to all documents in the textual collection. We used the Bisecting K-means [20,1] clustering algorithm to obtain topic hierarchies by taking into account the whole collection of documents. Note that this approach can only be adopted in controlled settings, but it is not feasible in practice, in particular for the incremental scenario we are investigating;
4. **Average Random Clustering (baseline):** represents the average result of several runs of a random clustering algorithm. We used this method to calculate a baseline for the F_{SCORE} measure.

For all methods, topic extraction is based on the selection of the most frequent terms of each cluster. Finally, it is important to note that during the execution of the incremental learning methods we maintained the same sampling S_0, the same subsets $S_1, ..., S_{10}$, and the same order of document insertions.

4.4   Retrieval Effectiveness of Topic Hierarchies

The results are presented and discussed considering two aspects: (1) model degradation during the incremental learning phase, and (2) quality of the topic hierarchies after the insertion of all documents.

Figure 1 shows the model degradation for during 10 different stages of the incremental updating. We have compared our BC^2 method (**Consensus+Incremental**) with Buckshot method (**UPGMA+Incremental**) in each document collection. To do so, the F_{SCORE} measure is calculated at the end of each stage, allowing to analyze the model degradation as the incremental clustering process takes place. In the graphs of Figure 1, the F_{SCORE} values of our method are shown with standard deviations obtained from 30 different trials. The **UPGMA+Incremental** is deterministic and thus only a single F_{SCORE} value is shown in the graphs. The experimental results illustrated in Figure 1 were statistically compared by using the nonparametric Wilcoxon matched-pairs signed-ranks test with a 95% confidence interval. One can observe that, our method is more robust with respect to model degradation. Thus, these results provide evidence that the use of consensus clustering to learning the initial model significantly improves the incremental learning of topic hierarchies. In particular, we highlight three desirable properties of the proposed method, namely:

1. **Low degradation rate:** the graphs from 20ng, ACM, Oshumed, and Reviews, illustrate scenarios in which the **Consensus+Incremental** obtains worse initial models compared to **UPGMA+Incremental**. However, the initial model obtained with our method indicates a low degradation rate of the model during the incremental learning, thereby making it capable of obtaining better final solutions;
2. **Stability:** the graphs from Hitech, NSF, PubMed-Cancer, Re8, and Reviews, illustrate the scenarios where the model obtained by the **Consensus+Incremental** is more stable during the incremental learning process compared to the traditional method, and
3. **Quality:** it can be seen from LATimes, Oshumed400, and WebACE, that both methods present some model degradation during the incremental updating. However, the topic hierarchies obtained with our method presents better quality in the overall process of incremental learning.

We also analyzed the quality of the topic hierarchies after the insertion of all documents. In this case, we compared topic hierarchies obtained by using an offline method (with access to all documents during the learning task). In principle, this method does not suffer from the effect of model degradation caused by incremental model updating. Figure 2 illustrates F_{SCORE} values obtained by each method (after inserting all documents). Recall that offline model is identified as "**Offline Clustering**", whereas "**Baseline**" identifies topic hierarchies obtained by using a random document clustering.

Statistical analysis of these results by using the Friedman and Post-hoc Nemenyi test (95% confidence level) reveals that (**Offline Clustering, Consensus+Incremental**) > **UPGMA+Incremental** > **Baseline**, where the order

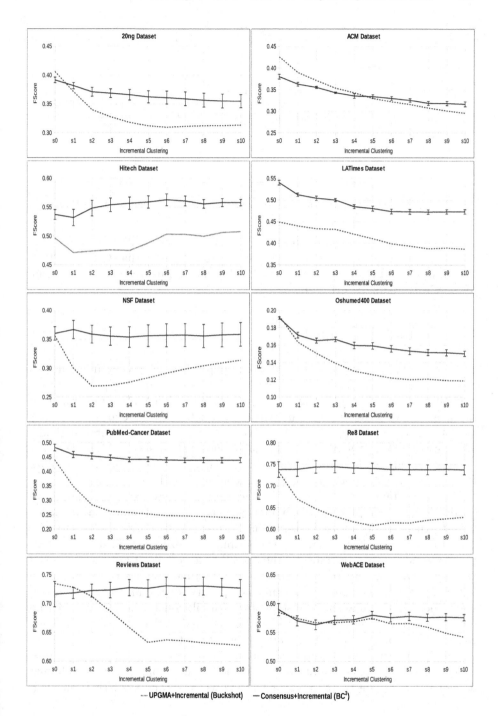

Fig. 1. Comparison of the methods during the incremental model updating

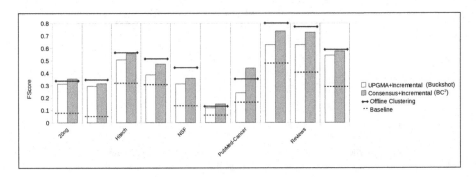

Fig. 2. Comparison of the method after the insertion of all documents

of the methods reflects the ranking position and ">" means that there are statistically significant differences. Although Figure 2 suggests that **Offline Clustering** provided better results, there is no statistical evidence that it is superior to the **Consensus+Incremental**.

5   Concluding Remarks

We presented a consensus clustering-based approach for incremental learning of topic hierarchies. The reported experimental results show that the proposed BC^2 method yields to more robust initial models, thus providing a lower degradation rates, as well as better stability and clustering quality. We have also shown that our BC^2 method is competitive even if compared to non-incremental methods. The investigation of the reasons that lead to such a surprising behavior is out of the scope of this paper, and it will be addressed in future works, but we speculate that incremental methods can be less susceptible to outliers.

It is important to recall that the hierarchical clustering algorithms have quadratic time complexity, which may present computational difficulties for large textual collections. However, the use of document sampling allows to mitigate this problem. For example, given a sample of size s obtained from a textual collection with n documents, the overall time complexity of the process is $O_{ini}(s)^2 + O_{inc}(nk)$. In this case, $O_{ini}(s)^2$ is the time complexity to obtain the initial model, and $O_{inc}(nk)$ is the time complexity for the incremental clustering, where k is the number of clusters of the initial model. Thus, by considering the dominant variables of the problem, and assuming that $s << n$, the time complexity turns out to be linear with respect to both n and k.

Finally, our approach is potentially useful for several real-world applications, which will be investigated in future works. For instance, approaches for including domain experts in the "exploratory data analysis loop" is particularly promising.

Acknowledgements. The authors wish to thank FAPESP for financial support (process 2010/20564-8 and 2011/19850-9).

References

1. Aggarwal, C.C., Zhai, C.: A survey of text clustering algorithms. In: Mining Text Data, pp. 77–128. Springer (2012)
2. Ayad, H., Kamel, M.S.: Topic Discovery from Text Using Aggregation of Different Clustering Methods. In: Cohen, R., Spencer, B. (eds.) Canadian AI 2002. LNCS (LNAI), vol. 2338, pp. 161–175. Springer, Heidelberg (2002)
3. Bouchachia, A.: Incremental learning. In: Encyclopedia of Data Warehousing and Mining, pp. 1006–1012. IGI Global (2008)
4. Carpineto, C., Osiński, S., Romano, G., Weiss, D.: A survey of web clustering engines. ACM Computing Surveys (CSUR) 41(3), 17:1–17:38 (2009)
5. Cutting, D.R., Karger, D.R., Pedersen, J.O., Tukey, J.W.: Scatter/gather: a cluster-based approach to browsing large document collections. In: 15th ACM SIGIR Int. Conf. on Research and Development in Inf. Retrieval, pp. 318–329 (1992)
6. Grossman, D.A., Frieder, O.: Information Retrieval: Algorithms and Heuristics. Springer, Secaucus (2004)
7. Guha, S., Rastogi, R., Shim, K.: Cure: an efficient clustering algorithm for large databases. ACM SIGMOD Record 27(2), 73–84 (1998)
8. Hofmann, T.: The cluster-abstraction model: unsupervised learning of topic hierarchies from text data. In: 16th IJCAI International Joint Conference on Artificial Intelligence, pp. 682–687 (1999)
9. Jain, A.K.: Data clustering: 50 years beyond k-means. Pattern Recognition Letters 31(8), 651–666 (2010)
10. Ke, W., Sugimoto, C.R., Mostafa, J.: Dynamicity vs. effectiveness: studying online clustering for scatter/gather. In: 32nd ACM SIGIR Int. Conf. on Research and Development in Inf. Retrieval, pp. 19–26 (2009)
11. Larsen, B., Aone, C.: Fast and effective text mining using linear-time document clustering. In: 5th ACM SIGKDD International Conference on Knowledge Discovery and Data Mining, pp. 16–22 (1999)
12. Liu, B.: Unsupervised Learning. In: Web Data Mining - Exploring Hyperlinks, Contents, and Usage Data, 2nd edn., pp. 133–166. Springer, Heidelberg (2011)
13. Marcacini, R.M., Rezende, S.O.: Incremental construction of topic hierarchies using hierarchical term clustering. In: 22nd International Conference on Software Engineering and Knowledge Engineering (SEKE), pp. 553–558. KSI Press (2010)
14. Muhr, M., Kern, R., Granitzer, M.: Analysis of structural relationships for hierarchical cluster labeling. In: 33rd ACM SIGIR Int. Conf. on Research and Development in Inf. Retrieval, pp. 178–185 (2010)
15. Pons-Porrata, A., Berlanga-Llavori, R., Ruiz-Shulcloper, J.: Topic discovery based on text mining techniques. Inf. Process. Manage. 43(3), 752–768 (2007)
16. Rokach, L.: A survey of clustering algorithms. In: Data Mining and Knowledge Discovery Handbook, 2nd edn., pp. 269–298. Springer (2010)
17. Sánchez, D., Moreno, A.: Creating Topic Hierarchies for Large Medical Libraries. In: Riaño, D., ten Teije, A., Miksch, S., Peleg, M. (eds.) KR4HC 2009. LNCS, vol. 5943, pp. 1–13. Springer, Heidelberg (2010)
18. Strehl, A., Ghosh, J.: Cluster ensembles - a knowledge reuse framework for combining multiple partitions. Journal of Machine Learning Research 3, 583–617 (2003)
19. Zavitsanos, E., Paliouras, G., Vouros, G.A.: Non-parametric estimation of topic hierarchies from texts with hierarchical dirichlet processes. Journal of Machine Learning Research 9, 2749–2775 (2011)
20. Zhao, Y., Karypis, G., Fayyad, U.: Hierarchical clustering algorithms for document datasets. Data Mining and Knowledge Discovery 10(2), 141–168 (2005)

Image Retrieval by Content Based on a Visual Attention Model and Genetic Algorithms

Milton Roberto Heinen[1] and Paulo Martins Engel[2]

[1] UDESC – Center of Technological Sciences
Joinville, CEP 89219-710, SC, Brazil
[2] UFRGS – Informatics Institute
Porto Alegre, CEP 91501-970, RS, Brazil
`miltonh@joinville.udesc.br, engel@inf.ufrgs.br`

Abstract. This paper proposes a new method for content-based image retrieval that uses a computational model of visual attention and genetic algorithm to find a given object in a set of images with different backgrounds. This method is composed by three main modules: a visual attention model that is quite robust against affine transformations; a color-based schematic representation of visual information; and a genetic algorithm that optimizes several parameters of the visual attention model in order to focus the attention mechanism on those regions of the image where it is most likely that a given object is present. The proposed method is validated through several experiments, and these experiments show that it can find the images that contain the sought object as well as the position and scale of the object in these images.

Keywords: Content-based image retrieval, visual attention, biomimetic vision, genetic algorithms, machine learning, computer vision, pattern recognition.

1   Introduction

Image collections have been created and used in several applications, such as digital libraries, medicine, and biodiversity information systems. Considering the increasing amount of image collections that results from popularization of digital cameras and the Internet, efficient search methods are becoming increasingly necessary. Content-based image retrieval systems aim to return the most relevant images in a database, according to the user's opinion for a given query [19]. Due to the dynamic nature of the problem, which may change the meaning of relevance among users for a same query, these systems usually rely on a learning process in which the user indicates a small set of images (training set) containing a sought object and the system returns the most relevant images to a given user's query [20].

In [12] a computational model of visual attention is used to identity groups of objects (i.e., clusters) in several images of the SIVAL image repository[1]. This work extends that previous approach by performing a content-based image retrieval based on the NLOOK visual attention model [10, 11, 13, 14], a color-based schematic representation of visual

---

[1] SIVAL Repository – `www.cs.wustl.edu/sg/accio/SIVAL.html`

L.N. Barros et al. (Eds.): SBIA 2012, LNAI 7589, pp. 122–131, 2012.

information and a genetic algorithm (GA). The NLOOK attention model was chosen because, as described in [14], it is less sensitive to 2D similarity transforms than other visual attention models as SAFE [4] NVT [15] and VOCUS [6]. Moreover, it can select the focus of attention (FOA) in a more accurate way than those attention models and can select scales of the FOAs as well as positions, too. However, NLOOK uses only bottom-up cues (visual scene elements that "pop-out") [16], which does not allow any kind of control over the regions selected as the focus of attention. To overcome this limitation, we use a genetic algorithm to evolve a set of parameters of the visual attention model in order to focus the attention mechanism on those regions of the image where it is most likely that a given object is present, thus implementing a top-down mechanism of attention [16].

This paper proposes a content-based image retrieval model that can find an object provided by the user in a set of images with the same object in different positions, light conditions and background. Moreover, it can also inform the position and scale of the object found on these images. More specifically, the user presents an image containing the object (the position and the size of the object in this image is also provided) and the system returns those images of the database that are most likely to contain the sought object and its corresponding localization and size in this image. In [21] an adaptive approach based on a neural network was used to drive the attention mechanism on some regions of the input image, but that approach uses only very simple, synthetic black and white images. The mechanism proposed in this paper, on the other hand, can work using real images of complex objects (e.g., fruits, cans, boxes) surrounded by different backgrounds and/or other objects.

The remaining of this paper is structured as follows: Section 2 describes the standard NLOOK visual attention model, which is the core of the mechanism proposed in this paper; Section 3 describes the proposed content-based image retrieval model, its characteristics and implementation details; Section 4 describes some experiments to demonstrate the performance of the proposed model; Finally, Section 5 provides some final conclusions and remarks.

2 NLOOK Visual Attention Model

Figure 1 shows the general architecture of the NLOOK visual attention model [14]. It is inspired by the scale-space concept [22], also used by Lowe on SIFT (*Scale Invariant Feature Transform*) descriptors [17]. Thus, unlike other attention models, which use just Gaussian pyramids, NLOOK uses scale-spaces to compute the linear center-surround operations, and this makes NLOOK less sensitive to 2D similarity transforms.

To create a scale-space, initially NLOOK sub-samples the input image into several octaves, and the initial image of each octave is the respective level of a Gaussian pyramid. Several scales are them created for each octave through the successive convolution of the initial image with Gaussian kernels. Finally, differences of Gaussians (DoG) are generated using the absolute difference of adjacent scale levels. Figure 2, adapted from [17], illustrates this process. NLOOK uses all possible octaves, i.e., it keeps generating octaves until the image is smaller than the lowest Gaussian kernel. This corresponds to five octaves to an 640×480 image and six octaves to an 1024×768 image.

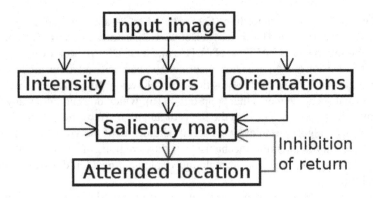

Fig. 1. General architecture of NLOOK

As recommended by Lowe [17], NLOOK generates three scales per octave with standard deviations of 1.2263, 1.5450 and 1.9466.

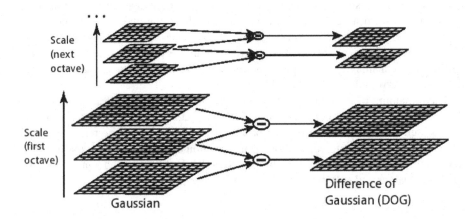

Fig. 2. DoG construction using scale-spaces

To generate the intensity maps $I(O, S)$ on NLOOK, where O and S are the number of octaves and scales, the source image is converted into a grayscale image I, and the DoG images of I are created using scale-spaces as described above. Thus, for a typical 1024×768 image, twelve DoG images (two per octave) are created. Each DoG image is normalized through the subtraction of its mean and the division by its standard deviation. To generate the color maps RG and BY, four scale-spaces $R(O, S)$, $G(O, S)$, $B(O, S)$ and $Y(O, S)$ are created from the (normalized by I) broadly-tuned color channels R, G, B and Y as described in [15]. Therefore, the absolute differences are computed between different channels, i.e., for an octave o DoG are computed through $|R(o, 0) - B(o, 1)|$ and $|R(o, 1) - B(o, 2)|$ for RG map and $|B(o, 0) - Y(o, 1)|$ and $|B(o, 1) - Y(o, 2)|$ for BY map.

The orientation maps $I(O, S, \theta)$ are created using Gabor filters, where θ is the pre-ferred orientation. These filters approximate the receptive field sensitivity profile of orientation selective neurons in primary visual cortex [2]. The orientation maps are cre-ated in a similar way as the intensity maps, but the grayscale pyramid $I(\varphi)$, used as initial image of octaves in intensity maps, is convolved with a Gabor filter before the scales generation. Four preferred orientations $\theta \in \{0°, 45°, 90°, 135°\}$ are used, gener-ating thus four scale-spaces of orientations. These four scale-spaces are normalized and combined into a unique scale-space of orientations.

After generating all feature scale-spaces, NLOOK combines them in a unique saliency scale-space $S(O, S)$ through the normalization and point-by-point addition of the corre-sponding octaves and scales. This scale-space $S(O, S)$ is then combined into a unique saliency map $\overline{S}$ by an expansion of every octave/scale to the original size (level 0), normalization and point-by-point addition. NLOOK creates the unique saliency map $\overline{S}$ at the finest level to avoid information losses. This unique saliency map $\overline{S}$ acts like a "summary" of $S(O, S)$, and also allows a unique inhibition or return (IOR) mechanism to be used.

After generating $S(O, S)$ and $\overline{S}$, NLOOK selects the FOAs in the following way: (i) a winner takes all (WTA) mechanism selects the most salient point of saliency map $\overline{S}$; (ii) $S(O, S)$ is analyzed to find out the octave/scale most important, i.e., the octave/scale with most effective contribution to the saliency at this point. If a draw occurs, NLOOK selects the finer octave/scale; (iii) the IOR mechanism, which has a spatial structure of an inverted Gaussian, is applied over $\overline{S}$. The diameter of IOR varies according to the most important octave/scale selected. More precisely, the standard deviation σ of the inverted Gaussian kernel is given by:

$$\sigma = \frac{P}{F^{o+s/(S-1)}} \tag{1}$$

where o is the most important octave, s is the most important scale, P is the patch size, which is a constant that defines the IOR base size (the default value is 17), S is the number of scales per octave (3 is the default value) and $F = 2^{-\frac{1}{S}}$ is the scale factor, that determines the IOR growing rate. Equation 1 allows large IOR sizes at coarser levels and small IOR sizes at finer levels. Next section describes how NLOOK is used in the proposed content-based image retrieval to find a given object in a set of images that may contain this object.

3   Image Retrieval by Content Using Visual Attention

As described in the previous section, NLOOK is able to find the most important features in a visual scene based on bottom-up cues, which makes it very suitable to some com-puter vision and robotic tasks such as object identification and landmark detection [7,8]. However, NLOOK does not allow any kind of control of the attention mechanism, i.e., the focus of attention is defined using only the information available in the input image. The biological attention mechanism present in the mammals, on the other hand, is mod-ulated by two main kinds of cues: bottom-up (visual scene elements that "pop-out") and top-down (information from brain that changes the attention focus) [16]. Therefore, a purely bottom-up attention mechanism does not allow the introduction of external information (e.g., heuristics) to drive the attention focus to those regions that are most similar to the user's query. In the human brain, for instance, the attention mechanism is modulated to become more sensitive to some kinds of input stimuli than others [3], and this modulation can be induced by learning [1].

In this paper a top-down mechanism, based on genetic algorithms, is used to drive the attention focus to those regions that are most similar to the user's query. This is accomplished by modulating the contribution of each feature map in the creation of the saliency scale-space $S(O, S)$ and the global saliency map $\overline{S}$. This modulation makes the attention mechanism more sensitive to some kinds of stimuli present in the user's query, thus allowing the localization of sought object in the current image without analyzing in details the entire visual scene (those regions containing the sought object will "pop-out").

To implement this modulation mechanism, a numeric value in the interval $[0, 1]$ is used to represent the contribution of each feature map in $S(O, S)$ and $\overline{S}$, i.e., each pixel of the corresponding map is multiplied by this value (which is evolved using the genetic algorithm) to adjust its relevance for the current user's query. Moreover, the high frequency components are attenuated using a Gaussian filter applied to each feature map, which allows the top-down mechanism to adjust the FOA size by evolving the standard deviation of the Gaussian filter. Therefore, the genetic algorithm has a genome composed by the following parameters:

- Contribution of each feature scale-space (intensity, colors and orientation) to the the saliency scale-space $S(O, S)$: 4 floating point parameters in the interval $[0, 1]$;
- Contribution of each octave in the global saliency map $\overline{S}$: 6 floating point parameters in the interval $[0, 1]$ (considering images of 1024×768 pixels);
- Contribution of each scale in the global saliency map $\overline{S}$: 3 floating point parameters in the interval $[0, 1]$;
- Standard deviation of the Gaussian filters: 4 floating point parameters (one for each feature scale-space) in the interval $[0, 3.5]$ (zero means that the scale-space will not be filtered);
- The patch size P (Equation 1): a single integer parameter in the interval $[1, 120]$. This parameter is used to change the size of the inhibition of return kernel and the corresponding FOA radius.

The genetic algorithm implementation used in the propotype is based on the GAlib[2] software library, developed by Matthew Wall of the Massachusetts Institute of Technology. GAlib was selected as it is one of the most complete, efficient and well known libraries for genetic algorithms simulation, and also it is a free open source library based on C++. A simple genetic algorithm proposed by Goldberg in his book [9] was used, and a floating point type genome was adopted. It used non-overlapping populations, elitism, roulette wheel selection schema and uniform crossover [9]. In order to reduce the search space, alleles were used to limit generated values to the intervals of each parameter described above. The fitness values were scaled using the sigma truncation scaling schema [18].

---

[2] GAlib – http://www.lancet.mit.edu/ga/

Based on the genetic algorithm described above, the proposed method can drive the focus of attention to those regions of the image that are more likely to contain the sought object. However, the fitness function needs to assess the level of quality of each solution, which is not so easy to compute in this kind of problem. To solve this problem we use the color angular indexing (CAI) [5] to codify the sought object and the current FOA, which allows to compute the fitness function very fast. According to [5], CAI is very compact representation scheme (only three numerical values) that is insensitive to illumination conditions, thus allowing us to identify a given object even at different scenarios, positions and scales. This representation scheme was also used in [12] to identify groups of objects of the SIVAL repository in an unsupervised way.

Therefore, the content-based image retrieval mechanism proposed in this paper works as follows:

1. The user presents some images containing the sought object and informs the position and radius of the circles encompassing the sought object in these images;
2. The CAI values of the sought object are computed in each image;
3. The genetic algorithm evolves the solutions in order to minimize the distance of the CAI values of the sought object and the main FOA of the current image;
4. After training the system analyses other images and returns those images that are likely to contain the sought object and the corresponding positions and scales of the object in each image as well.

The last step of this process requires the definition of a threshold TH in the interval $(0, \pi]$ to reject those images that do not contain the sought object (this threshold can also be evolved by the genetic algorithm). Next section presents some experiments to demonstrate the performance of the proposed model using real images of an apple at different positions, scales and illumination conditions.

4 Experiments and Results

This section describes the experiments to evaluate the performance of the proposed model using real images present at the SIVAL image repository. This repository contains 1500 images of 1024×768 pixels that depict 25 different objects (fruits, soda cans, books, boxes, clothes, etc.) shot at 60 different positions, lighting conditions and scenarios (on a carpet, on a chair, outdoors, in front of a white-board, etc) each. To perform this experiment we chose an apple as the sought object. A stratified 10-fold cross validation procedure was used to compute the results, and therefore each fold was composed by 6 images containing the sought object (an apple) and 6 images of the SIVAL repository that do not contain this object chosen at random, summing up 120 images. Table 1 shows the configuration parameters values used by the genetic algorithm in these experiments.

To assess the quality of the solutions, the position and scale of the apple on each image was measured and recorded in a text file. Thus the quality of the solutions was assessed by means of the Euclidean distance between the actual position and scale of the object present on the image and the position and scale of the main FOA. We have also performed the same experiment using the standard, bottom-up NLOOK to verify

Table 1. Parameters of the genetic algorithm

Parameter	Value
Crossover	0.80
Mutation	0.2
Population size	30
Number of generations	60

if our top-down mechanism is more efficient than the standard NLOOK. Table 2 shows the results obtained in these experiments.

Table 2. Results obtained in the experiments

	Standard NLOOK			Proposed approach		
	$\|x - x'\|$	$\|y - y'\|$	$\|r - r'\|$	$\|x - x'\|$	$\|y - y'\|$	$\|r - r'\|$
Mean	37.53	30.40	48.11	17.04	13.50	20.29
Std. dev.	78.26	35.42	24.54	14.86	14.13	19.51

The 2nd and 5th columns of Table 2 present the absolute differences between the x coordinates of the sought object and the main FOA computed using NLOOK and the proposed model, respectively. The 3rd and 6th columns present the absolute differences between the y coordinates of the sought object and the main FOA, and the 4th and 7th columns present the absolute differences between the measured radius of the object and the size of the main FOA. Figure 3 shows the some boxplot graphs computed over these experiments for the absolute difference between the x coordinates (Figure 3(a)), the y coordinates (Figure 3(b)) and the radius (3(c)), respectively.

We can notice that the proposed model really improved the global performance, because the distances between the desired and actual positions obtained using it are lower than the distances obtained using the standard (bottom-up only) NLOOK. Moreover, the proposed model is more precise in selecting the scale of the FOAs, thus selecting FOAs with sizes that generally encompass the target objects (the apple) in these images. Just to exemplify, Figures 4 and 5 show, respectively, the visual results obtaining using the standard NLOOK and the proposed model in some images chosen at random.

After computing the distances between the desired and actual positions, each image was analyzed to see if the main FOA is really over the sought object. Using the standard NLOOK the average error computed over the entire experiment was 8.73, i.e., in approximately 16% of the testing images (there are 54 images containing apples in the folds used for testing) the main FOA is outside the target object, whilst using the proposed model the average error was only 0.53, i.e., less than 1%. We have not computed the false positive rate using the standard NLOOK because it does not have a mechanism to reject images without the sought object. Using the proposed model, on the other hand, all these images were rejected by adjusting the threshold TH carefully.

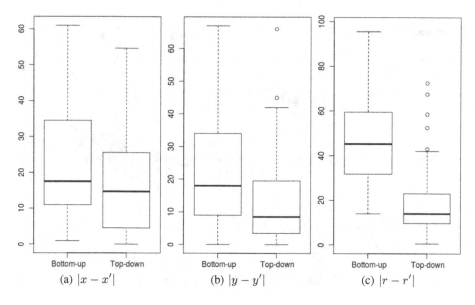

(a) $|x - x'|$ (b) $|y - y'|$ (c) $|r - r'|$

Fig. 3. Boxplot graphs of the results obtained in the experiments

Fig. 4. Some results obtained using the standard NLOOK (bottom-up cues only)

The time required to perform the evolution using six images for training is approximately five minutes in a typical computer[3]. Although this time is considable for a content-based image retrieval mechanism, it can be reduced by using images of small

---

[3] Dell Optiplex 755 computer, Intel(R) Core(TM)2 Duo CPU 2.33GHz processor, 1.95GB of RAM and SO Debian Linux 64 bits.

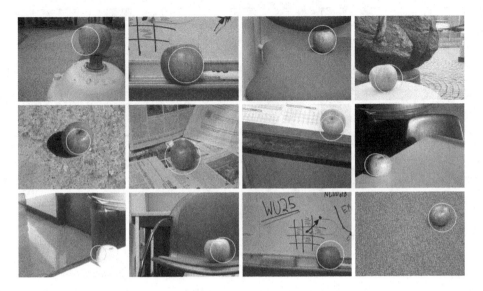

Fig. 5. Some results obtained using the proposed model (bottom-up and top-down cues)

dimensions (we have used 1024×768 images). In fact just those regions near the sought object can be preserved in the training images, and thus the evolution time can be significantly reduced.

5 Conclusion

This paper presents a content-based image retrieval mechanism, based on a computational model of visual attention, that can find the images that contain an object (the user's query) in a set of images that may contain or not this sought object. Moreover, it informs the position and scale of the object in the selected images, too. The content-based image retrieval mechanism is implemented using a genetic algorithm that adjusts some parameters of the attention model in order to change the focus of attention in regions that are similar to the sought object, thus allowing the use of top-down cues to drive the attention mechanism. The experiments described in this paper were aimed to verify if the FOAs computed by the attention model were close to the sought object, and thus we have not paid too much attention to those images that do not contain this object. Future perspectives include: (i) to perform experiments using other object of the SIVAL repository; (ii) to compute the rate of false positives produced by the model using a large number of images (e.g., all images of the SIVAL repository).

References

1. Connor, C.E., Egeth, H.E., Yantis, S.: Visual attention: Bottom-up versus top-down. Current Biology 14(19), 850–852 (2004)
2. Daugman, J.G.: Complete discrete 2-d gabor transforms by neural networks for image analysis and compression. IEEE Trans. Acoustics, Speech, and Signal Processing 36(7), 1169–1179 (1988)
3. Desimone, R., Duncan, J.: Neural mechanisms of selective visual attention. Annual Reviews Neuroscience 18(1), 193–222 (1995)
4. Draper, B.A., Lionelle, A.: Evaluation of selective attention under similarity transformations. Computer Vision and Image Understanding 100(1), 152–171 (2005)
5. Finlayson, G.D., Chatterjee, S.S., Funt, B.V.: Color Angular Indexing. In: Buxton, B.F., Cipolla, R. (eds.) ECCV 1996. LNCS, vol. 1065, pp. 16–27. Springer, Heidelberg (1996)
6. Frintrop, S.: VOCUS: A Visual Attention System for Object Detection and Goal-directed Search. Ph.D. dissertation, Rheinische Friedrich-Wilhelms-Universitat Bonn, Bonn, Germany (January 2006)
7. Frintrop, S., Jensfelt, P.: Attentional landmarks and active gaze control for visual SLAM. IEEE Trans. Robotics, Special Issue on Visual SLAM 24(5) (October 2008)
8. Fritz, G., Seifert, C., Paletta, L., Bischof, H.: Attentive Object Detection Using an Information Theoretic Saliency Measure. In: Paletta, L., Tsotsos, J.K., Rome, E., Humphreys, G.W. (eds.) WAPCV 2004. LNCS, vol. 3368, pp. 29–41. Springer, Heidelberg (2005)
9. Goldberg, D.E.: Genetic Algorithms in Search, Optimization and Machine Learning. Addison-Wesley, Reading (1989)
10. Heinen, M.R., Engel, P.M.: Avaliação de modelos de atenção visual em relação a transformações afins. In: Proc. IV Workshop de Visão Computacional (WVC 2008), Bauru, SP, Brazil (November 2008) (text in Portuguese)
11. Heinen, M.R., Engel, P.M.: NLOOK: Modelo de atenção visual relativamente insensível a transformações afins. Hífen 32(62), 270–277 (2008) (text in Portuguese)
12. Heinen, M.R., Engel, P.M.: Categorização de objetos utilizando atenção visual. In: Anais do IX Congr. Brasileiro de Redes Neurais / Inteligência Computacional (CBRN). p. 5. Sociedade Brasileira de Redes Neurais (SBRN), Ouro Preto, MG, Brazil (October 2009) (text in Portuguese)
13. Heinen, M.R., Engel, P.M.: Evaluation of visual attention models under 2d similarity transformations. In: Proc. 24th ACM Symposium on Applied Computing (SAC 2009) – Special Track on Intelligent Robotic Systems, pp. 1156–1160. ACM press, Honolulu (2009)
14. Heinen, M.R., Engel, P.M.: NLOOK: A computational attention model for robot vision. Journal of the Brazilian Computer Society (JBCS), 3–17 (September 2009)
15. Itti, L., Koch, C., Niebur, E.: A model of saliency-based visual attention for rapid scene analysis. IEEE Trans. Pattern Analysis and Machine Intelligence 20(11), 1254–1259 (1998)
16. Lee, K.W., Buxton, H., Jianfeng, F.: Cue-guided search: A computational model of selective attention. IEEE Trans. Neural Networks 16(4), 910–924 (2005)
17. Lowe, D.G.: Distinctive image features from scale-invariant keypoints. Int. Journal of Computer Vision 60(2), 91–110 (2004)
18. Mitchell, M.: An Introduction to Genetic Algorithms. The MIT Press, Cambridge (1996)
19. da Silva, A.T., Ao, A.X.F., Aes, L.P.M.: Active learning paradigms for cbir systems based on optimum-path forest classification. Pattern Recognition 44(12), 2971–2978 (2011)
20. da Silva Torres, R., Ao, A.X.F.: Content-based image retrieval: Theory and applications. Journal of Theoretical and Applied Informatics (RITA) 13(2), 161–185 (2006)
21. Van de Laar, P., Heskes, T., Gielen, S.: Task-dependent learning attention. Neural Networks 10(6), 981–992 (1997)
22. Witkin, A.P.: Scale-space filtering. In: Proc. Int. Joint Conf. Artificial Intelligence, pp. 1019–1022. Morgan Kaufman, Karlsruhe (1983)

A Symbolic Representation Method to Preserve the Characteristic Slope of Time Series

Willian Zalewski[1,2], Fabiano Silva[1], Feng Chung Wu[2],
Huei Diana Lee[2], and André Gustavo Maletzke[2]

[1] Federal University of Parana – UFPR, Curitiba, Brazil
Formal Methods and Artificial Intelligence Laboratory – LIAMF
{wzalewski,fabiano}@inf.ufpr.br
http://www.inf.ufpr.br
[2] State University of West Parana – UNIOESTE, Foz do Iguassu, Brazil
Bioinformatics Laboratory – LABI
{andregustavom,hueidianalee,wufengchung}@gmail.com
http://www.foz.unioeste.br/labi

Abstract. In recent years many studies have been proposed for knowledge discovery in time series. Most methods use some technique to transform raw data into another representation. Symbolic representations approaches have shown effectiveness in speedup processing and noise removal. The current most commonly used algorithm is the Symbolic Aggregate Approximation (SAX). However, SAX doesn't preserve the slope information of the time series segments because it uses only the Piecewise Aggregate Approximation for dimensionality reduction. In this paper, we present a symbolic representation method to dimensionality reduction and discretization that preserves the behavior of slope characteristics of the time series segments. The proposed method was compared with the SAX algorithm using artificial and real datasets with 1-nearest-neighbor classification. Experimental results demonstrate the method effectiveness to reduce the error rates of time series classification and to keep the slope information in the symbolic representation.

Keywords: Time Series, Knowledge Discovery, Symbolic Representation, Classification, Dimensionality Reduction.

1 Introduction

Time Series Data Mining (TSDM) has attracted enormous attention in the last decade motived by the increase of stored data and the interest in the discovery of relevant hidden informations. One important problem to the knowledge discovery process is treating data with temporal dependencies [1–4]. In particular, the time series are collections of observations made chronologically and this type of data is present in almost all domains such as business, industry, medicine, science and entertainment [5].

The inherent characteristics of time series data includes: large datasets, high dimensionality, continuous updates, high feature correlation and the unavoidable

L.N. Barros et al. (Eds.): SBIA 2012, LNAI 7589, pp. 132–141, 2012.

noise [1, 2, 5]. The performance of the TSDM algorithms degrades with regard to high dimensionality [6]. Therefore, this kind of data needs a different treatment from that used for no time related attributes [4]. In the last decade, methods and techniques have been proposed to transform raw time series data in other representations. Specifically, symbolic representations have demonstred to be a very effective tool to reduce the dimensionality of the time series [2, 6–8]. In the data mining context, the symbolic representations should preserve the underlying information and produce interpretable symbols within the domain [4, 9].

In this work, we proposed a symbolic representation method to preserve the characteristic slope between the time series observations. Our approach is based on a combination of the basic ideas presented in [10] and [11].

The rest of this paper is organized as follows. Section 2 presents background on time series data mining and related works. Section 3 introduces our symbolic method. Section 4 contains an experimental evaluation of the symbolic method on a variety of time series datasets. In Section 5 the effectiveness of the symbolic method is also discussed. Finally, Section 6 presents the conclusions and directions for future works.

2   Background and Related Works

In this section we provide some preliminary definitions about time series used in this work and present a brief review on time series data minig.

2.1   Definitions

Definition 1. *Time Series:* A time series $T = \{t_1, \ldots, t_m\}$ is an ordered collection of m real values.

Definition 2. *Segment:* Given a time series $T = \{t_1, \ldots, t_m\}$ a segment S of T is a collection of n contiguous values of T, that is $S = \{t_i, \ldots, t_{i+n}\}$, for $1 \leq i \leq m - n$.

Definition 3. *Symbolic Time Series:* Given a time series $T = \{t_1, \ldots, t_m\}$ and a finite alphabet a, the symbolic time series $\hat{T}$, obtained from T by applying a discretization function f, can be represented as a sequence of symbols $f(T, a) = \hat{T} = \{\hat{t}_i, \ldots, \hat{t}_n\}$, for $n \leq m$.

2.2   Time Series Data Mining

The data mining field has attracted a lot of attention during the last years since it involves methods and algorithms capable of automatically discovering some new nontrivial patterns or unexpected relationships using a combination of techniques from machine learning, mathematics, statistics and database technologies [3] [5].

Traditional data mining methods were developed to analyze data without temporal relation. However, the storage increase of continuous data with temporal interdependencies, such as time series, has motived the development of new data mining approaches [4].

TSDM is a relatively new area that uses data mining methods ajusted to take into consideration the temporal nature of data [5]. Over the last decade many interesting TSDM techniques were proposed and have shown to be useful in many applications [3].

In the context of TSDM, the time series representation is a fundamental problem because direct manipulation of high dimensional data in an efficient way is extremely difficult in traditional data mining techniques. A common approach is to use a time series representation based on some dimensionality reduction technique, while preserving the relevant characteristics of a particular dataset [1, 4, 8].

2.3   Time Series Representations

Many numerical time series representation approaches have been proposed in the literature to reduce the high dimensionality, such as the Discrete Fourier Transform (DFT); the Discrete Wavelet Transform (DWT); Singular Value Decomposition (SVD); Discrete Cosine Transform (DCT); Chebchevy Polinomials (CP); Piecewise Linear Approximation (PLA); Piecewise Aggregate Approximation (PAA); Adaptative Piecewise Constant Approximation (APCA); and approaches based on Perceptually Important Points (PIPs) identification [1, 2, 5, 8, 12].

There are domains, such as medicine and finances, where symbolic representation rather than numerical analisys is needed to produce more comprehensive knowledge of the time series [13]. Therefore, many works have also considered symbolic representations of time series, such as Shape Description Alphabet (SDA) [14]; Interactive Matching of Patterns with Advanced Constraints in Time Series Databases (IMPACTS) [15]; Clustering equal-sized segments by sliding window [10]; Application of self-organizing maps in the first order differences between consecutive raw data values [11]; *Clipping* [16]; *Persist* [9]; and Piecewise Vector Quantized Approximation (PVQA) [17].

Most of the symbolic representations cited are affected by two main aspects. Firstly, the intrinsic dimensionality of the symbolic representation is the same as the raw data, thus the data mining algorithms scale poorly with high dimensionality. Second, the unavoidable noise presence in time series can produce meaningless symbols. The Symbolic Aggregate Approximation (SAX) is a symbolic approach that applies dimensionality reduction technique as a preprocessing step, in this case the PAA algorithm [18].

Piecewise Aggregate Approximation: To transform m-dimension to w-dimension, the data is divided into w equal-size segments, and the mean value of each segment is used to represent original time series with lower w-dimension. The time series $T = \{t_1, \dots, t_m\}$ of length m can be represented in w-dimensinal space by a vector $\overline{T} = \{\overline{t}_1, \dots, \overline{t}_w\}$ and the ith element of $\overline{T}$ is calculated by the Equation 1 [18]:

$$\overline{t}_i = \frac{w}{m} \sum_{j=\frac{m}{w}(i-1)+1}^{\frac{m}{w}i} t_j \tag{1}$$

Symbolic Aggregate Approximation: The SAX symbolic representation is performed in two steps. Firstly, PAA algorithm is applied to the raw time series to dimensionality reduction (Figure 1(a)). Second, the distribution space (y-axis) is divided into equiprobable regions under a Gaussian curve and the mean segment values from PAA are converted into symbols corresponding to each segment [18]. The SAX symbolic representation transformation can be defined by the function $SAX(\overline{T}, w, a) = \hat{T} = \{\hat{t}_1, \ldots, \hat{t}_w\}$ where $\hat{t}_i$ represent the ith symbol, w is the number of segments and a is the alphabet size. In the Figure 1(b) is presented a SAX example of a symbolic sequence **baabccbc** with the alphabet $\{a, b, c\}$.

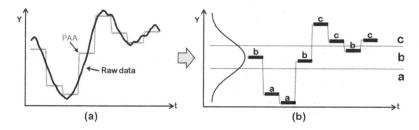

Fig. 1. (a) PAA application example and (b) SAX application example

In the last decade, SAX has been widely applied to many fields and obtained very good results [1, 2, 12]. However, the smoothing effect by only using the PAA algorithm may lose useful information, especially the segment slope information. Futhermore, the equiprobable feature of SAX symbols produces low performance for non-uniform time series [7].

3   The Proposed Symbolic Representation Method

In this section we present a new symbolic representation method for time series. The method is performed in three sequential steps: (1) Dimensionality Reduction; (2) Data Transformation; and (3) Symbolic Representation.

3.1   Dimensionality Reduction

Dimensionality reduction step is the simple PAA application. PAA algorithm is used to convert a time series of length m to w segments in reduced time series $\overline{T} = \{\bar{t}_1, \ldots, \bar{t}_w\}$ according to Equation 1. The mean values from PAA representation are viewed as points into the new reduced space.

3.2   Data Transformation

In this work, we proposed a intermediate step between dimensionality reduction and symbolic representation. Data transformation step is based on the basic idea presented in [11] which calculates the first order differences instead of the use the raw time series data. But, in our approach a dimensionality reduction technique is applied before. Thus, the calculus of the first order differences is performed on the reduced space $\overline{T}$ produced by PAA algorithm.

For each pair of adjacent elements $(i, i + 1)$ in reduced time series $\overline{T}$, where $1 \leq i \leq w - 1$, the new first order difference value is $\delta(i) = \overline{t}_{i+1} - \overline{t}_i$ and the transformed time series $\delta(\overline{T})$ is given by $\delta(\overline{T}) = \{\delta(i), \ldots, \delta(w - 1)\}$.

In the Figure 2 we provide an example to visualize the result of the data transformation step.

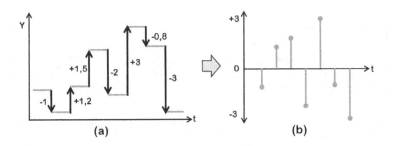

Fig. 2. (a) PAA first order differences and (b) First order differences transformation

3.3   Symbolic Representation

The idea behind our symbolic representation approach is based in [10], which applies clustering algorithm on time series segments to discretization. The main difference between this previous work to our method is the dimensionality reduction in preprocessing step as suggested in SAX approach. Furthermore, our method try to keep the slope information by the application of the data transformation step.

Symbolic transformation is perfomed based on time series produced by the data transformation step. Firstly, a k-means technique with Lloyd algorithm is used to separe in k groups all $\delta(\overline{T})$ values. The k value represents the alphabet size for symbolic representation. After, the k clusters centers $C = \{c_1, .., c_k\}$ (Figure 3(a)) are used to transform $\delta(\overline{T})$ into symbols (Figure 3(b)). In this approach, the input to the clustering algorithm can be only one time series or a set of time series in the same domain.

The symbolic transformation is performed by a function called $Symb$ (Equation 2) that receives as input a $\delta(i)$ value and the clusters centers $\{c_1, .., c_k\}$ and computes the correspondent symbol.

$$Symb(\delta(i), C) = which.min(\{|c_j - \delta(i)|, \ldots, |c_k - \delta(i)|\}) \qquad (2)$$

where the function $which.min$ finds the c_j value that has the minimum difference to the $\delta(i)$ value.

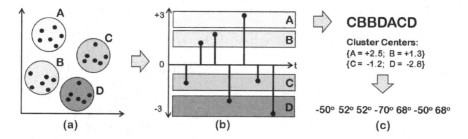

Fig. 3. (a) k-means application; (b) symbolic transformation and (c) symbolic representation

The function $Symb$ is applied for each value in $\{\delta(i), \ldots, \delta(w-1)\}$ and the result set is the symbolic representation $\hat{T}$ of the time series T. In the Figure 3(c) is presented an example of the resultant symbolic representation **CBBDACD** considering the alphabet size $a = 4$.

The values of symbols represent the difference information between the points in $\overline{T}$ representation. Thus, it is possible to associate to the symbols (cluster centers) one more meaningful information, such as the slope angle which is given by $\alpha = tan^{-1}(c_j)$. In the example of Figure 3(c) the symbolic sequence **CBB-DACD** can be represented by the angles values $(-50^o, +52^o, +52^o, -70^o, +68^o, -50^o, +68^o)$. Furthermore, our method finds the angle division boundaries based on data characteristics domain.

4   Experimental Validation and Results

In this section we present the experimental evaluation to validate the effectiveness of our proposed symbolic representation. In our experiments we use 20 time series datasets provided by the UCR Time Series Data Mining Archive [19] that contains artificial and real-world data. Some dataset features are presented in Table 1, such as dataset name (1st column), number of classes (2nd column), size of training set (3rd column), size of testing set (4th column) and time series length (5th column).

In this work, we provide a initial comparison with SAX method by a experimental classification using one nearest neighbor classifier and Euclidean Distance as similarity measure between two symbolic sequences. The method codification was built using R Language.

For a particular dataset, all time series within its training set are used to create the clusters centers. The k-means algorithm was applied with maximum number of 1000 random starts and maximum number of 1000 iterations. We performed experiments on different combinations of dimensionality reduction and alphabet size for each dataset. We evaluated the alphabet size a in the interval from 2 until 20 of time series length and the dimensionality w in the interval from 2 until 50% of time series length. Each time we increase by two the value of w.

Table 1. Datasets details and experimental 1-NN classification results

Name Dataset	Number Classes	Size Train	Size Test	Length Series	Error EU	Error SAX	Error Symb	w/a SAX	w/a Symb
Synthetic Control	6	300	300	60	0.120	**0.013**	0.080	12/13	8/8
Gun-Point	2	50	150	150	0.087	0.180	**0.033**	6/19	72/13
CBF	3	30	900	128	0.148	0.083	**0.050**	10/14	6/5
FaceAll	14	560	1690	131	**0.286**	0.299	0.317	36/14	32/11
OSU Leaf	6	200	242	427	0.483	0.467	**0.425**	128/10	24/13
Swedish Leaf	15	500	625	128	0.211	0.235	**0.156**	42/18	38/18
50words	50	450	455	270	**0.369**	0.397	0.391	16/7	40/14
Trace	4	100	100	275	0.240	0.270	**0.160**	44/16	58/7
Two Patterns	4	1000	4000	128	0.093	**0.063**	0.182	64/10	16/5
Wafer	2	1000	6174	152	0.0045	0.0034	**0.0027**	26/4	20/5
FaceFour	4	24	88	350	0.216	0.170	**0.090**	128/10	24/5
Lightning2	2	60	61	637	0.246	0.213	**0.163**	256/10	26/5
Lightning7	7	70	73	319	0.425	**0.342**	0.342	6/11	8/8
ECG	2	100	100	96	0.120	0.100	**0.040**	32/7	24/16
Adiac	37	390	391	176	0.389	0.836	**0.299**	80/19	54/15
Yoga	2	300	3000	426	0.170	0.178	**0.163**	192/16	64/8
Fish	7	175	175	463	0.217	0.308	**0.114**	104/15	72/6
Beef	5	30	30	470	0.467	0.433	**0.066**	28/16	190/19
Coffee	2	28	28	286	0.250	0.107	**0.000**	132/19	14/7
Olive Oil	4	30	30	570	0.133	0.833	**0.030**	4/2	228/15

The results are shown in Table 1. Error rates for Euclidean Distance, SAX and our approach are presented in the 6th, 7th and 8th columns, respectively. Parameters values w and a for SAX and our method are presented in the 9th and 10th columns, respectively. For both the presented results of the SAX approach and of our method, the w and a selected parameters correspond to the best error rate classification obtained.

In the Figure 4 we summarize the results by plotting the error rates and the dimensionality reduction for each dataset into pairwise scatter plots. The points below the diagonal line indicate that our symbolic representation method is more accurate (Figure 4(a)) or has a greater dimensionality reduction than the SAX approach (Figure 4(b)).

In the statistical evaluation of the symbolic representations performance, we use the approach applied in [22]. The Iman and Davenport version of the F-test is used to test the null-hypothesis that all symbolic representations have the same performance and the observed differences are merely random. As post hoc test we used the Bonferroni-Dunn test.

In our accuracy analysis the corresponding critical value is equal to 9.04 for $\alpha = 0.05$ (mean ranks: SAX=2.27, Symb=1.32, EU=2.40). The null-hypothesis that all methods has the same accuracy is rejected (p-value is 0.0003). In the post hoc test we verified significantly difference between Symb and SAX, and Symb and EU.

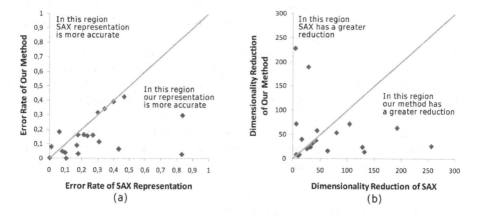

Fig. 4. (a) Error rate comparison (b) Dimensionality reduction comparison

5 Discussion

Time series data mining techniques have become an important tool to discover novel relevant patterns that can help in decision making process. The human decision making in time series analysis is commonly based on domain expert perceptions [20]. In this cases, a symbolic representation is prefered instead a numerical representation [13] and the symbols should preserve the underlying information [6].

The SAX is a symbolic representation approach, based on PAA, that has been widely used in the literature [1, 2, 7]. Dimensionality reduction by PAA has advantages such as fast processing and smoothing the noise. However, this approach causes a high possibility to miss important patterns in time series data, such as the local trend of the time series [12, 21]. Futhermore, the Gaussian assumption of the symbols distribution has effects on the SAX performance for non-uniform or correlated time series and for dimensionality reduction [7].

In our symbolic representation method, we also use dimensionality reduction by PAA algorithm to smooth the noise. But, our approach preserves part of the local trend due to transformation step which computes the first order differences between the PAA resulting values. Therefore, the representation of the differences between segments also embodies the whole trend of the time series.

An important contribution of our method is the meaningful symbols creation based on the slope characteristic of the time series segments in a particular domain. This feature is obtained by the k-means algorithm application in a training set. Other relevant characteristic inherent to our method is the possibility to partially rebuild the original signal from the symbolic representation using the angle values of each symbol.

The statistical evaluation indicates that our method is more accurate than SAX and EU for one nearest neighbor classification. In the most datasets evaluated our method was more accurate, for example, the accurate in the datasets has an improvement higher than 50% in comparison with SAX. In special, the error rate for *Olive Oil* dataset was reduced from 83.3% to 3% and for *Beef* dataset the error rate was reduced from 46.4% to 6.6%. Furthermore, a significative dimensionality reduction was obtained for most datasets.

According with the chart in the (Figure 4(a)) in only three datasets SAX outperforms our method. In these datasets our method has difficulty in treating the time series which contain noisy segments separating the most relevant segments and also in treating the noisy time series with constant trend.

Our method have one disadvantage in comparison with SAX, the requirement of a training set to create the clusters centers in a previous step. However, it is a good symbolic representation alternative to preserve the local slope information.

6   Conclusions and Future Works

In this paper we have presented a symbolic representation method to preserve the slope information among the time series segments. We have performed a evaluation on 20 widely used datasets including artifical and real-world time series. The experimental results analysis demonstrate the efectiveness of our representation method in time series classification for low error rates and for dimensionality reduction in comparison with SAX approach.

Future works include the application of other techniques to improve the symbolic boundaries search and for dimensionality reduction, such as APCA. Other tasks are introduce a lower bounding distance measure to our method and evaluate it using other classification algorithms.

Acknowledgments. We would like to acknowledge Dr. Eamonn Keogh for his experimental datasets.

References

1. Fu, T.C.: A review on time series data mining. Engineering Applications of Artificial Intelligence 24, 164–181 (2010)
2. Lin, J., Keogh, E., Wei, L., Lonardi, S.: Experiencing SAX: a novel symbolic representation of time series. Data Mining and Knowledge Discovery 15, 107–144 (2007)
3. Laxman, S., Sastry, P.S.: A survey of temporal data mining. Sadhana 31, 173–198 (2006)
4. Antunes, C.M., Oliveira, A.L.: Temporal Data Mining: an overview. In: KDD Workshop on Temporal Data Mining, pp. 1–13 (2001)
5. Karamitopoulos, L., Evagelidis, G.: Current Trends in Time Series Representation. In: Proceedings of 11th Panhellenic Conference on Informatics, pp. 217–226 (2007)
6. Hugueney, B.: Adaptive Segmentation-Based Symbolic Representations of Time Series for Better Modeling and Lower Bounding Distance Measures. In: Fürnkranz, J., Scheffer, T., Spiliopoulou, M. (eds.) PKDD 2006. LNCS (LNAI), vol. 4213, pp. 545–552. Springer, Heidelberg (2006)

7. Pham, N.D., Le, Q.L., Dang, T.K.: Two Novel Adaptive Symbolic Representations for Similarity Search in Time Series Databases. In: Conference International AsiaPacific Web, pp. 181–187 (2010)
8. Ding, H., Trajcevski, G., Scheuermann, P., Wang, X., Keogh, E.: Querying and mining of time series data: experimental comparison of representations and distance measures. Time 1, 1542–1552 (2008)
9. Morchen, F., Ultsch, A.: Optimizing time series discretization for knowledge discovery. In: Proceeding of the Eleventh ACM SIGKDD International Conference on Knowledge Discovery in Data Mining, KDD 2005, pp. 660–665 (2005)
10. Das, G., Ip Lin, K., Mannila, H., Renganathan, G., Smyth, P.: Rule discovery from time series, pp. 16–22. AAAI Press (1998)
11. Giles, C.L., Lawrence, S., Tsoi, A.C.: Noisy Time Series Prediction using a Recurrent Neural Network and Grammatical Inference. Neural Networks 44, 161–183 (2001)
12. Li, H., Guo, C., Qiu, W.: Similarity measure based on piecewise linear approximation and derivative dynamic time warping for time series mining. Expert Systems with Applications (2011)
13. Alonso, F., Martínez, L., Pérez, A., Santamaría, A., Valente, J.P.: Modelling Medical Time Series Using Grammar-Guided Genetic Programming. In: Perner, P. (ed.) ICDM 2008. LNCS (LNAI), vol. 5077, pp. 32–46. Springer, Heidelberg (2008)
14. Andre-Jonsson, H., Badal, D.Z.: Using Signature Files for Querying Time-Series Data. In: Komorowski, J., Żytkow, J.M. (eds.) PKDD 1997. LNCS, vol. 1263, pp. 211–220. Springer, Heidelberg (1997)
15. Huang, Y.W., Yu, P.S.: Adaptive query processing for time-series data. In: Proceedings of the Fifth ACM SIGKDD international Conference on Knowledge Discovery and Data Mining, KDD 1999, pp. 282–286 (1999)
16. Bagnall, A.J., Janacek, G.J.: Clustering time series from arma models with clipped data. In: Proceedings of the Tenth ACM SIGKDD International Conference on Knowledge Discovery and Data Mining, KDD 2004, pp. 49–58. ACM (2004)
17. Megalooikonomou, V., Faloutsos, C.: A Multiresolution Symbolic Representation of Time Series. In: 21st International Conference on Data Engineering ICDE 2005, pp. 668–679 (2005)
18. Lin, J., Keogh, E., Lonardi, S., Chiu, B.: A symbolic representation of time series, with implications for streaming algorithms. In: Proceedings of the 8th ACM SIGMOD Workshop on Research Issues in Data Mining and Knowledge Discovery DMKD 2003, pp. 2–11 (2003)
19. Keogh, E., Zhu, Q., Hu, B., Hao. Y., Xi, X., Wei, L., Ratanamahatana, C.A.: The UCR Time Series Classification/Clustering Homepage (2011), www.cs.ucr.edu/~eamonn/time_series_data/
20. Batyrshin, I., Sheremetov, L.: Perception-based approach to time series data mining. Applied Soft Computing 8, 1211–1221 (2008)
21. Lkhagva, B., Suzuki, Y., Kawagoe, K.: Time Series Representation ESAX for Financial Applications. In: Proceedings of the 22nd International Conference on Data Engineering Workshops, p. 115 (2006)
22. Gorecki, T., Luczak, M.: Using derivatives in time series classification. Data Mining and Knowledge Discovery, 1–22 (2012)

Orchestrating Multiagent Learning of Penalty Games

Ana L.C. Bazzan

PPGC / Instituto de Informática, Universidade Federal do Rio Grande do Sul (UFRGS)
Caixa Postal 15.064 – 91.501-970, Porto Alegre, RS, Brazil
bazzan@inf.ufrgs.br

Abstract. In comparison to single agent learning, reinforcement learning in a multiagent scenario is more challenging, since there is an increase in the space of combination of actions that may have to be explored before agents learn an efficient policy. Among other approaches, there has been a proposition to address this problem by means of biasing the exploration. We follow this track using an organizational structure where low-level agents mainly use reinforcement learning, while also getting recommendations from agents possessing a broader view. These agents keep a base of cases in order to give such recommendations, orchestrating the process. We show that this approach is able to accelerate and improve learning in penalty games, a especial case of coordination games.

1 Introduction and Related Work

The problems posed by a high number of agents in a multi-agent reinforcement learning (MARL) scenario are well known. These problems arise mainly due to the fact that while one agent is trying to model the environment (other agents included), the others are doing the same and potentially changing the environment they share. Besides, in MARL an issue is the exponential increase in the space of joint states and joint actions. Although in the meantime there are different interpretations of what is a joint state and/or action, here this means that an agent i *explicitly models the states and actions of the other agents*. In fact, in simple Markov decision processes (MDP), a reward is associated with a tuple $\langle s, a, s' \rangle$ denoting the fact that i was in state s, performed action a and ended up in s' with reward r. Contrarily, in multiagent MDPs (also known as stochastic games or SGs), the reward of i is determined as $r^i : S^1 \ldots \times S^k \times A^1 \times \ldots \times A^k \rightarrow \mathbb{R}$, where k denotes each agent in the set. The decision on whether or not to include joint states and/or joint actions in the learning process of i is a key one as it has severe implications. In fact, most of the game-theoretic literature concentrates on games with few players and few actions because otherwise it is computationally prohibitive.

In this paper we address the stochastic version of a common-interest game (a subclass of coordination games (CGs), where all agents have the same payoff). In particular, we use penalty games.

In [2], three kinds of CGs are used to investigate what the authors call individual learners (IL) and joint-action learners (JAL). Although they deal with repeated stage games (one state), action selection is stochastic. Thus the convergence to a good coordinated equilibrium is not guaranteed. After this work, multiagent learning in repeated and stochastic games were addressed by [3, 6, 7, 8, 10], among others. However, these

L.N. Barros et al. (Eds.): SBIA 2012, LNAI 7589, pp. 142–151, 2012.

works mostly deal with two or three agents at most, and/or assume perfect monitoring (observation of joint actions) by all agents at least in the beginning of the time periods. Regarding other forms of SGs (i.e., not necessarily common-interest games), the zero-sum case was discussed in [9] and attempts of generalizations to general-sum SG appeared in [5] (Nash-Q).

Returning to the focus of this paper, three approaches that are indeed based on some form of non-explicit biased exploration are discussed next. The first is due to Hines and Larson [4] who use repeated (not stochastic) games where agents can follow the advice of a mediator that makes suggestions to the agents as to what actions to take. A similar approach to the one in the present paper appears in [1], which, however, considers not only repeated games but also SGs. The third form of biased exploration appears in [11] where a supervision framework (MASPA) is introduced to speed up the convergence of MARL. Hierarchically superior agents keep abstract states of lower-level agents. This view is used to generate rules or suggestions, passed down to local agents. The authors argue that the learning process can be coordinated using some kind of organizational control for biasing the exploration; otherwise it would take too long for a high number of agents.

The approach followed in the present paper differs from the previously mentioned works as follows. Regarding MARL in repeated and stochastic games in general, it relaxes the assumption of perfect monitoring at the level of agents present in many of them. Rather, it delegates part of this monitoring to an organizational control. Regarding approaches that are based on biased exploration ([4, 11]), in the present paper communication is restricted to agents one level up or down the hierarchy, thus reducing significantly the amount of exchanged messages. Finally, it can be extended to a high number of agents. Note that this has not being the case in the works mentioned (except [11]) when explicit modeling of joint actions is involved.

In the present paper, the focus is on the repeated penalty game mentioned in [2] because here convergence is not always smooth due to the high miscoordination penalty and because this game has multiple optimal joint actions, thus making the coordination task harder than in the scenarios tackled in [11].

The next section introduces CGs by means of the so-called penalty game. Sections 3 and 4 present the proposed approach and its evaluation respectively. Conclusions and future directions appear in Section 5.

2 Learning in Coordination Games

In its simplest version, a CG is a type of matrix game. Actions are selected according to a pure strategy (one that puts probability one in a single action), or according to a mixed strategy (there is a probability distribution over the available actions). All games have at least one equilibrium, possibly in mixed strategies. Some kinds of CGs have clearly more than one equilibrium, thus the selection of an efficient joint action – the coordination task – is not trivial.

Nowadays CGs are of great interest to model for instance the establishment of standard for innovative technologies (e.g., music/video players, etc.), or for coordination tasks in general (e.g., mobile robots sharing common paths, etc.). An example of a CG

Table 1. Payoff-matrix for a penalty game ($k < 0$)

		Agent 1		
		a_0	a_1	a_2
	b_0	10	0	k
Agent 2	b_1	0	2	0
	b_2	k	0	10

is the penalty game, whose payoff matrix is shown in Table 1. The variable $k < 0$ is the penalty for miscoordination (in case two agents select particular different actions). This game has three deterministic equilibria, of which two ($< a_0, b_0 >, < a_2, b_2 >$) are preferred. When k approaches zero the attractiveness of action 1 is not significant. However, for a high penalty, say $k = -100$, both agents find actions 0 and 2 less attractive and end up selecting the "safe" action (action 1). In [2], it is shown that the probability of convergence to one of the optimal equilibria is influenced by the magnitude of k: roughly (this depends on other parameters), for $-40 < k < 0$ convergence is achieved, while for higher negative penalties this may not be the case.

3 Improving Coordination in Penalty Games

This section details the proposed approach, which is based on a hierarchical organization of agents. At the bottom level, so called low-level agents interact (here playing a penalty game) and learn using Q-learning (Eq. 1) in an independent way (i.e., Q values are computed only for individual actions). These low-level agents are grouped (based on a given criterion) so that each group has a supervisor that observes its agents for a time period, records *joint actions* and their rewards in a base of cases, where a case is the set of rewards agents get and the solution of this case is the joint action. After this period is over, the supervisor looks for the best reward it has observed so far and recommends a joint action. Thus, supervisors have a broader view, even if it is not detailed.

$$Q(s, a) \leftarrow Q(s, a) + \alpha \ (r + \gamma \ max_{a'} \ Q(s', a') - Q(s, a)) \tag{1}$$

Here our approach differs from the one by Zhang *et al.*. As mentioned, in their case, hierarchically superior agents keep abstract views of lower-level agents, which are used to generate either rules (that agents must follow) or suggestions (that are optional), passed down to local agents. These rules may be conflicting as for instance when they forbid all actions. Thus we have opted to follow a different approach: supervisors recommend actions instead of forbidding them.

3.1 Formal Setting

An n-agent repeated game is a tuple (N, A, R) where N, A, R are the set of agents, the discrete action space (set of joint actions), and the reward function. For the specific

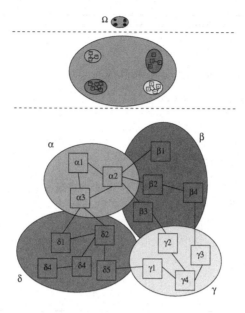

Fig. 1. Example of a three-level organization: 16 agents $(\alpha_1, ..., \beta_1, ..., \delta_4)$ in the lower level, supervised by α, β, γ, and δ in the second level; the top-level supervisor is Ω

case discussed here, the set of actions and respective payoffs are as in the payoff matrix (Table 1) with variable $k < 0$.

Contrarily to previous works, we let agents play a penalty game repeatedly with $m > 2$ other agents. As illustration we refer to Figure 1 in which low-level agents are divided in groups $(\alpha_1, ..., \alpha_3, \beta_1, ..., \beta_4, \gamma_1, ..., \gamma_5, \delta_1, ..., \delta_4)$ that are supervised by 4 supervisors $(\alpha, \beta, \gamma, \delta)$, which in turn is supervised by supervisor Ω. In this figure, a line indicates that the two agents sharing it play the CG. This however does not mean that they communicate explicitly. Groups in the figure can be thought as departments in an organization, each having a supervisor. Although interactions mainly occur among agents within their departments or groups, some agents also have interactions outside their groups (as, e.g., α_2). The task of coordination could be relatively simple if each group were closed. In this case, they could learn a policy that would allow them to coordinate over one of the actions. However, since the groups are not closed, the interactions with outsiders make this task more complicated because a cascade of coordination tasks exists. For example, all agents in the α group only learn an efficient action selection (say, 0) if agents in the β group do the same; otherwise agent α_2 would have a low payoff given that it would get a penalty for miscoordinating either with α_1 and α_3, or with β_1, β_2 and β_3. Ideally α_2 and all its neighbors $(\alpha_1, \alpha_3, \beta_1, \beta_2, \beta_3)$ all select the same action between the two preferred ones, so that α_2 could receive the total payoff of 50 (10 from each of the 5 plays). This implies that all neighbors of α_2's neighbors should select the same action and so on.

Algorithm 1. Individual learning stage (stage 1)

1: **while** $t \leq \Delta_{ind}$ **do**
2:    **for all** $L_j \in \mathcal{L}$ **do**
3:       when in state s_j, select action a_j with probability $\dfrac{\exp^{Q(s_j,a_j)/T}}{\sum_{a_j \in A_j} \exp^{Q_{s_j,a_j}/T}}$ (Boltzmann exploration) and observe reward
4:       update Q_j^{ind} {// Eq. 1 }
5:    **end for**
6:    **for all** $S_i \in \mathcal{S}$ **do**
7:       observe state, action, and reward for each L_j
8:       compute the average reward $\bar{r}$ (among L_j's)
9:       **if** tuple $< \vec{a}^t, \vec{s}^t, \bar{r} >$ not yet in the base of cases **then**
10:         add tuple $< \vec{a}^t, \vec{s}^t, \bar{r} >$
11:       **else**
12:         **if** $\bar{r} > \overline{r_{old}}$ **then**
13:           replace by tuple $< \vec{a}^t, \vec{s}^t, \bar{r} >$
14:         **end if**
15:       **end if**
16:    **end for**
17: **end while**

3.2 Individual and Supervised Learning

The period in which low-level agents interact and learn by using Q-learning is called stage 1, which lasts for Δ_{ind} time steps. Each agent updates its Q table considering the rewards received by playing with m interacting neighbors.

As discussed previously, individual learning is not necessarily efficient in penalty games, especially when the absolute value of k increases. Hence our approach is to use supervisors to orchestrate the process giving recommendations to agents they supervise; Supervisors do not actually play the game thus they are not included in the set N of low-level agents. In fact, supervisors must be seen as facilitators or tutors that will observe the local agents' in their groups from a broader perspective and recommend actions to them. This recommendation is based on a group perspective, in opposition to the purely local perspective of low-level agents. The supervised learning works as in algorithms 1 to 3 and explained next. We remark that, although the case-study in this paper is a repeated penalty game (thus single state), these algorithms are general in the sense that a case may include the joint states as well as joint actions. Hence, in algorithms 1 to 3 a case is actually the tuple $< \vec{a}^t, \vec{s}^t, \bar{r} >$. Regarding scalability, the only issue may be the number of *supervised* agents because the higher this number, the higher the combination of states and actions that has to be observed.

The main parameters of the model are: the learning rate α, the discount rate γ, and the threshold τ (explained below); the set of low-level agents $N = \mathcal{L} = \{L_1, ..., L_n\}$; the set $\mathcal{S} = \{S_1, ...\}$ of supervisor agents; Δ_{ind} (time period during which each L_j learns and acts independently, updating the Q table Q_j^{ind}); Δ_{tut} (time period during which each S_i prescribes an action to each L_j in its group based on cases observed so far); and Δ_{crit} (time period during which each L_j can act independently or follow the

---

Algorithm 2. Tutoring stage (stage 2)

1: **while** $\Delta_{ind} < t \leq \Delta_{ind} + \Delta_{tut}$ **do**
2:    **for all** $S_i \in \mathcal{S}$ **do**
3:       communicate with supervisor at upper level; get similar cases; add to case base
4:       given $\vec{s}^t$, find $\vec{a}^t$ in case base for which $\bar{r}$ is highest; communicate a_j to each L_j
5:    **end for**
6:    **for all** $L_j \in \mathcal{L}$ **do**
7:       perform action a communicated by supervisor, collect reward {or follow local policy if supervisor has not prescribed any action}
8:       update Q_j^{ind} {// Eq. 1 }
9:    **end for**
10:   **for all** $S_i \in \mathcal{S}$ **do**
11:      observe state, action, and reward for each L_j
12:      compute the average reward (among L_j's) $\bar{r}$
13:      **if** tuple $< \vec{a}^t, \vec{s}^t, \bar{r} >$ not yet in case base **then**
14:         add tuple $< \vec{a}^t, \vec{s}^t, \bar{r} >$
15:      **else**
16:         **if** $\bar{r} > \overline{r_{old}}$ **then**
17:            replace by tuple $< \vec{a}^t, \vec{s}^t, \bar{r} >$
18:         **end if**
19:      **end if**
20:   **end for**
21: **end while**

---

recommendation of the supervisor). These time periods are henceforth called stages 1 to 3 respectively.

Stage 1 is described in Algorithm 1. Each L_j learns a policy; each supervisor S_i observes its low-level agents and records information to a base of cases. Thus this base is composed by the tuples mentioned before where $\bar{r}$ is *averaged* over all supervised agents. The case that has yielded the highest $\bar{r}$ so far is kept in the base (line 12).

The second stage takes further Δ_{tut} time steps. In this stage, each S_i: i) observes the joint state of its low-level agents; ii) communicates with its upper-level supervisor; iii) gets similar cases; iv) updates the base of cases with the best case received (if any); v) retrieves $\vec{a}^t$ for which $\bar{r}$ is the highest. It is important to note that in any case the local Q tables continue to be updated. The main difference to stage 1 is that at stage 2, low-level agents are committed to the action prescribed by the supervisor, even when the expected reward is not as good as the computed Q values.

In the third stage (which takes Δ_{crit} steps) low-level agents need not follow the prescribed action. Rather, after comparing the expected reward (r^e) that was communicated by the supervisor, with the locally computed Q value for this particular prescribed action, each agent may select the action associated with its local policy. This means that the low-level agent will only select the prescribed action if this is at least as good as the expected Q value (here considering a tolerance factor τ as in line 7 in Algorithm 3). This reinforces the autonomous character of a low-level agent, thus the approach is not centralized. No matter whether the low-level agents do follow the prescriptions or not, supervisors are able to observe actions and rewards and update their bases of cases.

Algorithm 3. Critique stage (stage 3)

1: **while** $\Delta_{ind} + \Delta_{tut} < t \leq \Delta_{ind} + \Delta_{tut} + \Delta_{crit}$ **do**
2:    **for all** $S_i \in \mathcal{S}$ **do**
3:        given $\vec{s}^t$, find $\vec{a}^t$ in case base for which $\bar{r}$ is maximal; communicate a_j^p to each L_J plus expected reward $r^e = \bar{r}$
4:    **end for**
5:    **for all** $L_j \in \mathcal{L}$ **do**
6:        {//compare Q_j^{ind} and r^e:}
7:        **if** $r^e \times (1 + \tau) > Q_j^{ind}$ **then**
8:            perform a_j^p {//where a_j^p is action prescribed by supervisor for this agent}
9:            update Q_j^{ind}
10:       **else**
11:           perform a_j^{ind} {// where a_j^{ind} is selected according to local policy}
12:           update Q_j^{ind} {// Eq. 1 }
13:       **end if**
14:    **end for**
15:    **for all** $S_i \in \mathcal{S}$ **do**
16:        observe state, action, and reward for each L_j
17:        compute the average reward (among L_j's) $\bar{r}$
18:        **if** tuple $< \vec{a}^t, \vec{s}^t, \bar{r} >$ not yet in case base **then**
19:            add tuple $< \vec{a}^t, \vec{s}^t, \bar{r} >$
20:        **else**
21:            **if** $\bar{r} > \overline{r_{old}}$ **then**
22:                replace by tuple $< \vec{a}^t, \vec{s}^t, \bar{r} >$
23:            **end if**
24:        **end if**
25:    **end for**
26: **end while**

---

4   Empirical Results

4.1   Settings

Experiments were performed using the penalty game discussed previously with agents located in a grid. In particular we have used 4×4, 8×8, 16×16, and 24×24 but we only report the latter. Others follow the same trend. Joint actions and their rewards are as in Table 1, thus the highest possible average reward is 10.

To render the explanation and experiments more clear, let us assume that the agents form an organization, where the relationship is characterized by the physical position in a non-toroidal grid. Hence low-level agents interact with the 4 nearest neighbors, not all belonging to the same group. As explained, this makes the coordination task not trivial. Agents play the game repeatedly with their neighbors while rewards are recorded.

Values for the main parameters used in the simulation are given in Table 2 (unless otherwise said), where the symbol $*$ means that the value varied. Plots of average reward (over all agents) along time are shown. All experiments were repeated at least 100 times. To keep the figures more clear error bars are not shown; however the standard deviation is 10% at most.

Table 2. Parameters and their values

Parameter	Description	Value	Parameter	Description	Value
$N = \|\mathcal{L}\|$	number of agents	*	$k_\bullet$	penalty	*
T	temperature	16	T decay	temperature decay	$0.99 \times T$
α	learning coefficient	0.5	γ	discount rate	0.0
τ	intolerance factor	*	Δ_{ind}	stage 1	200
Δ_{tut}	stage 2	10	Δ_{crit}	stage 3	190

4.2 Experiments

In order to verify the conclusions in [2], we have performed experiments without exploration and with individual learning only. We have not used JAL here as it proved inefficient (as also verified in [10]). After, we have performed experiments using both individual learning and supervised learning.

Initially, to prove that exploration is key in CGs, Figure 2 (gray line) depicts the performance when no exploration is used (here for $k = -150$; however the trend to inefficient learning does not change regarding other values). The low performance is caused by the fact that, at each time step, we can expect one-third of the agents to select one of the three actions, thus leading to high miscoordination. Figure 2 also shows the reward along time for individual learning with exploration, for different values of k. With low penalty (e.g., -50) individual learning is not as inefficient as when penalty is -100 or -150. We have performed simulations with various grid sizes and note that for $k = -150$, no matter the grid size, reward converges to approximately 2 (reward of action 1), showing that individual learning is not efficient in the penalty game (agents could get 10), justifying the need for more efficient learning approaches, such as the one proposed here, when the penalty is high. In stage 1, local agents act individually thus we can expect performance to be similar to the situation discussed above.

In Figure 3 we plot both the individual and the supervised learning for $k = -100$ (dotted and black curves respectively). One can see that the recommendation pays off. In stage 2 (starting at time 200), there is an increase in the reward due to good cases communicated by the higher level of supervision. In stage 3, which starts at time 210, low-level agents may refuse to do the action prescribed by the supervisor. In order to decide whether or not to refuse, a low-level agent compares the reward the supervisor is expecting with the value of its Q table. The plot in Figure 3 refers to $\tau = -0.9$ meaning that one agent accepts a recommendation only in case its Q-value is lower than 10% of $\bar{r}$. In this (quite extreme) situation many recommendations are not accepted in the beginning yielding a bad reward, as shown by the drop of reward value in Figure 3 around time 210. Agents that do follow the recommendation tend to get a good reward. With this, their supervisors collect better cases, which also "pull" the other agents they supervise towards a coordinated action. The final reward (see Figure 3) is $r_{400} = 8.25$, which is 78% higher compared with $r_{400} = 4.71$ for the individual learning.

We have tried other values for τ as well. When $\tau \geq 0$ the learning process may take longer as agents accept more recommendations and therefore explore less, which in turn prevents the supervisors from observing different cases. Also, plots similar to Figure 3 were generated for other values of k. We omit them here but note that the final rewards

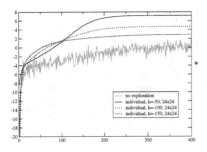
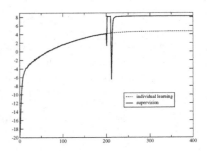

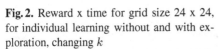

Fig. 2. Reward x time for grid size 24 x 24, for individual learning without and with exploration, changing k

Fig. 3. Reward along time for grid size 24 x 24: individual learning; supervision (black line); $k = -100$

when $k = -150$ are: $r_{400} = 2.89$ (individual learning) and $r_{400} = 8.17$ (supervision), proving that the latter is robust to higher penalties.

Obviously, with $24x24$ agents, it is not always the case that all groups of agents coordinate over either action 0 or action 2 in all 100 repetitions of the same situation. Clusters of agents may sometimes select one or another action. Therefore it is difficult to get the average reward close to 10, unless k is very low. Interestingly, agents in the border of such clusters end up selecting action 1 thus avoiding the penalty. These agents form a "wall" between groups selecting different actions.

4.3 Discussion

Next, some issues are discussed, which are important regarding the employability of such approach. It is possible to observe that individual learners, if left alone would converge to the expected payoff only when the penalty is not high. By using recommendation, agents immediately receive new information about rewards that are possible but that they have not yet seen due the size of state-action space.

The orchestration by the supervisors of course means additional communication when compared to individual learning. Now agents have to exchange two messages each with the respective supervisor. However, other approaches rely on even more communication as they are based on observation of joint actions by every agent. Moreover, due to privacy issues, it might not be reasonable to expect agents to reveal their actions to others, especially if these belong to different groups.

5   Conclusion

Multi-agent reinforcement learning is inherently more complex than single agent reinforcement learning, thus a MDP-based formalism when applied to games with many agents is not necessarily efficient. This paper proposes an approach based on supervision as a compromise between complete distribution and complete centralization. Supervisors have a broader view and are able to recommend more efficient joint actions. We have measured the reward with and without supervision, and the results show that the former pays off in penalty games, where it is known that the coordination task is hard.

One possible extension for the current work refers to groups of different sizes, in which case it is necessary to process the cases that refer to sets of agents with different cardinality. Besides, the currently fixed allocation of supervisors to low-level agents may be relaxed so that the supervisor could be re-allocated on the basis of the quality of its recommendation.

References

[1] Bazzan, A.L.C.: Coordinating many agents in stochastic games. In: Proc. of the IEEE IJCNN 2012 (June 2012)

[2] Claus, C., Boutilier, C.: The dynamics of reinforcement learning in cooperative multiagent systems. In: Proceedings of the Fifteenth National Conference on Artificial Intelligence, pp. 746–752 (1998)

[3] Guestrin, C., Lagoudakis, M.G., Parr, R.: Coordinated reinforcement learning. In: Proceedings of the Nineteenth International Conference on Machine Learning (ICML), pp. 227–234. Morgan Kaufmann, San Francisco (2002)

[4] Hines, G., Larson, K.: Learning when to take advice: A statistical test for achieving a correlated equilibrium. In: McAllester, D.A., Myllymäki, P. (eds.) UAI, pp. 274–281. AUAI Press (2008)

[5] Hu, J., Wellman, M.P.: Multiagent reinforcement learning: Theoretical framework and an algorithm. In: Proc. 15th International Conf. on Machine Learning, pp. 242–250. Morgan Kaufmann (1998)

[6] Kapetanakis, S., Kudenko, D.: Reinforcement learning of coordination in cooperative multi-agent systems. In: AAAI/IAAI, pp. 326–331 (2002)

[7] Kuminov, D., Tennenholtz, M.: As safe as it gets: Near-optimal learning in multi-stage games with imperfect monitoring. In: Proceeding of the ECAI 2008, pp. 438–442. IOS Press, Amsterdam (2008)

[8] Lauer, M., Riedmiller, M.: An algorithm for distributed reinforcement learning in cooperative multi-agent systems. In: Proc. 17th International Conference on Machine Learning, pp. 535–542. Morgan Kaufmann, San Francisco (2000)

[9] Littman, M.L.: Markov games as a framework for multi-agent reinforcement learning. In: Proceedings of the 11th International Conference on Machine Learning, ML, pp. 157–163. Morgan Kaufmann, New Brunswick (1994)

[10] Wang, X., Sandholm, T.: Reinforcement learning to play an optimal nash equilibrium in team markov games. In: Advances in Neural Information Processing Systems 15, NIPS 2002 (2002)

[11] Zhang, C., Abdallah, S., Lesser, V.: Integrating organizational control into multi-agent learning. In: Sichman, J.S., Decker, K.S., Sierra, C., Castelfranchi, C. (eds.) Proceedings of the 8th International Conference on Autonomous Agents and Multiagent Systems (AAMAS), Budapest, Hungary (2009)

An Architectural Model
for Autonomous Normative Agents

Baldoino F. dos Santos Neto[1], Viviane Torres da Silva[2], and Carlos J.P. de Lucena[1]

[1] PUC-Rio, Informatics Department, LES, Rio de Janeiro, Brazil
[2] Universidade Federal Fluminense, Computer Science Department, Niterói, Brazil
{bneto,lucena}@inf.puc-rio.br, viviane.silva@ic.uff.br

Abstract. Social norms have become one of the most promising approaches that use an informal social control to ensure a desirable social order in open multi-agent systems. In these systems, autonomous and heterogeneous entities work towards similar or different goals. Norms regulate the behaviour of agents by defining obligations and prohibitions, and by creating rewards and penalties to encourage the agents to meet theses constraints. The development of autonomous normative agents, i. e., agents able to take decisions by following their motivations related to their goals and taking into account the system norms, has proven to be significantly more challenging than the design of traditional agents. In this paper, we introduce an architectural model that provides a set of functions to support the agent on the reasoning about the norms. These operations assist the agent to perceive the system's norms, detect the fulfillment and violation of the norms while verifying their activation and deactivation, select the norms the agent intends to fulfill, identify and overcome conflicts among norms.

1 Introduction

In order to regulate the behavior of agents and ensure a desirable social order, normative mechanisms have been used. Such mechanisms establish a set of norms that regulate the behaviour of the agents by defining obligations and prohibitions. Moreover, norms may give stimulus to their fulfillment by defining rewards and may discourage their violation while stating punishments [8].

Although norms are promising mechanisms to regulate agents' behavior, agents are autonomous. Therefore, there might be circumstances in which they prefer to violate a norm and reach a private goal than to fulfill the norm.

With the aim to cope with this issue, the use of *Autonomous Normative Agents*(ANA) has been encouraged [3]. ANA are autonomous BDI (*Belief-Desire-Intention*) agents [8] able to join a society of agents, adopt the system norms and deal with them in an autonomous way. They are able to reason about their motivations while taking into account the existence of social norms and to determine whether to follow a norm and receive its rewards, or violate it and receive the corresponding punishment. Although being considered a promising approach, the development of ANA agents has shown to be significantly more challenging than traditional BDI agents. The normative reasoning process of ANA agents must consider several issues related to norms that are not considered in the traditional BDI reasoning process, such as: (i) the decision making related to the adoption

L.N. Barros et al. (Eds.): SBIA 2012, LNAI 7589, pp. 152–161, 2012.

of norms; (ii) the verification of their fulfillment and violation; (iii) the detection and overcome of conflicts among norms at runtime; (iv) the selection of the norms to be fulfilled and the ones to be violated; and (v) the influence of norms on the generation of new goals or update of the agent's motivations. Although it is possible to find in the literature a wide range of available solutions to build agents able to deal with motivations and norms (e.g., [3][8] [6]), none of them present a complete solution to model motivation-driven entities that take into account the norms but are not driven by them.

In this context, the paper presents an architecture to model motivation-oriented agents able to deal with the norms of a society in an autonomous way. The architecture is specified by using the Z specification language [7] and extends BDI model by including norms related functions to support normative reasoning. An agent built according to the proposed architecture has a function to check the incoming perceptions (including norms), to adopt new norms addressed to the agent, to detect the fulfillment or violation of the norms and to check the activation and deactivation conditions. Moreover, the architecture also defines a norm selection function that is responsible for three tasks: to identify and overcome conflicts among the conflicting norms, to select the norms the agent intends to fulfill (based on their motivations), and to influence the agent to take place the decisions for fulfilling or violating a norm by generating new goals or updating their motivations. We illustrates our approach by describing a social commerce [1] system that has a norm-reinforcement mechanism and agents playing the *Social Network Analyst (SNA)* role. Social networking sites, such as Facebook, due to their popularity, have become an environment tremendously interesting to apply different strategies related to social commerce. Although social commerce involves several aspects, as discussed in [1], we focus on two: (**selling**) that uses social media to assist in the on line selling of products; and (**promoting**) that promotes advertisements and referrals of products.

2   Norms and Autonomous Normative Agents Structure

In this section we describe and explain the characteristics and structure of the autonomous normative agents and of the norms being considered in this work.

2.1   Norm Structure

```
┌─ Norm ──────────────────────────────────────
│  1 : addressees : ℙ₁ Addressee
│  2 : deonticConcept : DeonticConcept
│  3 : activation, deactivation : Context
│  4 : state : State
│  5 : rewards, punishments : ℙ State
└──────────────────────────────────────────────
```

In this paper, as represented in the norm schema, we adopt the idea presented in [8], where norms specify environment states (line 4) that must be achieved by an agent (its obligations OBL) or states that are prohibited to be achieved (its prohibitions PRO) (line 2). Moreover, we consider, as discussed in [8], that a norm specifies the set of agents that are directly responsible for fulfilling it (line 1). Since a norm is not applied all the time, but only in particular circumstances, a norm also specifies the situations in which the agents must fulfill it (line 3). As discussed [8], in order to discourage the agents to violate the norms and promote their fulfillment, norms may include rewards to be provided when they are fulfilled and punishments when violated (line 5).

The norms used by the norm-reinforcement mechanism of the social commerce system are: **Norm-1**: *(Addreesse)*: SNA agents; *(Deontic Concept)*: OBL; *(Activation Condition)*: a new product is launched; *(Deactivation Condition)*: 15 days after the launch; *(State)*: SNA has 15 days to promote the product; *(Reward)*: SNA's reputation is increased; *(Punishment)*: SNA's reputation is decreased; **Norm-2** *(Addressee)*: SNA agents; *(Deontic Concept)*: OBL; *(Activation Condition)*: a new product is launched; *(Deactivation Condition)*: 15 days after the launch; *(State)*: SNA is obligated to sell at least 10 products in 15 days; *(Reward)*: SNA's reputation is increased; *(Punishment)*: SNA's reputation is decreased; **Norm-3** *(Addressee)*: SNA agents; *(Deontic Concept)*: PRO; *(Activation Condition)*: the promotion rate of a *SNA* agent related to any product is lower than 2 promotions per month; *(Deactivation Condition)*: the promotion rate of a *SNA* agent related to any product is greater than 2 promotions per month; *(State)*: SNA is prohibited to sell; *(Punishments)*: SNA's reputation is decreased;

2.2   Autonomous Normative Agent (ANA) Structure

___AutonomousNormativeAgent_____

1 : *self* : *AgentName*
2 : *role* : *Role*
3 : *beliefs* : $\mathbb{P}_1$ *Belief*
4 : *goals* : $\mathbb{P}_1$ *Goal*
5 : *intentions* : $\mathbb{P}_1$ *Intention*
6 : *norms* : $\mathbb{P}$ *Norm*
7 : *goalmotivation* : *Goal* $\rightarrow \mathbb{Z}$
8 : *normativestate* : *Norm* $\rightarrow$ *NormativeState*
9 : *normativesatisfaction* : *Norm* $\rightarrow$ *NormativeSatisfaction*
10 : *normativeselection* : *Norm* $\rightarrow$ *NormativeSelection*

NormativeState ::= *adopted* | *activated* | *deactivated*
NormativeSatisfaction ::= *fulfilled* | *violated* | *not_analyzed*
NormativeSelection ::= *tobe_fulfilled* | *tobe_violated* | *not_selected*

A ANA agent, as described in the *AutonomousNormativeAgent* schema[1], is an entity that has a name (line 1); is playing a role in the system (line 2); has a set of beliefs (line 3); can achieve a set of goals (line 4); has a set of intentions (line 5); and is the addressee of a set of norms (line 6). As discussed in [8] and represented by the relation *goalmotivation* (line 7), each goal (i. e., the environment state that the agent has the intention to achieve) is associated with a motivation that drives the behaviour of the agent. Motivations are preferences represented by a (positive or negative) value.

---

[1] It is an extension of the autonomous agent structure[8].

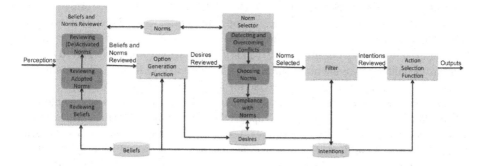

Fig. 1. Autonomous Normative Agent Architectural Model

A positive motivation related to a goal indicates that the agent wants to achieve the state and a negative that it does not want to achieve it.

In order to monitor the current normative situation of a norm (i. e., if the norm is adopted, activated, deactivated, fulfilled or violated) the agent store information about the norms by using the following relations (line 8, 9 and 10): (*normativestate*) indicates if the norm is adopted, activated or deactivated (see *NormativeState* type); (*normativesatisfaction*) indicates if the norm was fulfilled or violated (see *NormativeSatisfaction* type); and, (*normativeselection*) indicates if the norm was selected to be fulfilled or violated (see *NormativeSelection* type).

3 Architectural Model

Our architectural model extends BDI to help agents on reasoning about the system norms that are considered to be a primary concept that influences the agents in all their decision making. The extensions are represented in the architecture (Figure 1) by the following components: *Beliefs and Norms Reviewer*, *Norms Selector* and *Norms*.

3.1 Beliefs and Norms Reviewer

Besides reviewing the beliefs (represented by the *Reviewing Beliefs* task in Figure 1), this function helps the agent to recognize its responsibilities towards other agents through two tasks:

ReviewAdoptedNorms
1 : $\Delta NormativeAgent$
2 : $norms? : \mathbb{P} \, Norm$

3 : $(\forall \, norm : norms? \mid norm \notin norms \, \wedge$
4 : $\quad (\, self \in addresseeAgentName \, norm.addressees \, \vee$
5 : $\quad role \in addresseeRole \, norm.addressees) \, \bullet$
6 : $\quad\quad norms' = norms \cup \{norm\}$
$\quad\quad normativestate = normativestate \oplus \{(norm, adopted)\})$

Reviewing Adopted Norms: The *Reviewing Adopted Norms* task, formalized in schema *ReviewAdoptedNorms*, receives as input the norms (line 2) and reviews the adopted norms (line 6) by checking (as already described in [8]): *(i)* if the new norm does not exist in the set of norms (line 3); and *(ii)* if the agent (or the role played by the agent) is the addressee of the norm (lines 4 and 5, respectively).In the social commerce system, all the three norms are addressed to the agent playing the role *SNA*.

Reviewing (De)Activated Norms: The main goals of this task, as represented in Schema *ReviewDeActivatedNorms*, is to update the states of the norms that have became active (line 1); fulfilled or violated (lines 2 and 3); and inative (lines 4 and 5). An obligation norm is fulfilled if it is active and its state is achieved (line 2), and it is violated if it is inactive and the state has not been achieved (line 5). A prohibited norm is violated if it is active and its state is achieved (line 3), and it is fulfilled if it is inactive and the state has not been achieved (line 4).

ReviewDeActivatedNorms
ReviewAdoptedNorms

$\forall\, norm : norms \bullet$

1 : $(\,(\,normativestate\ norm = adopted \lor normativestate\ norm = deactivated\,) \land$
   $\quad norm.activation \subseteq beliefs \Rightarrow$
   $\quad normativestate = normativestate \oplus \{(norm, activated)\}\,) \land$

2 : $(\,norm.deonticconcept = OBL \land normativestate\ norm = activated \land$
   $\quad norm.state \subseteq beliefs \Rightarrow$
   $\quad normativesatisfaction = normativesatisfaction \oplus \{(norm, fulfilled)\}) \land$

3 : $(\,norm.deonticconcept = PROH \land normativestate\ norm = activated \land$
   $\quad norm.state \subseteq beliefs \Rightarrow$
   $\quad normativesatisfaction = normativesatisfaction \oplus \{(norm, violated)\}) \land$

4 : $(\,norm.deonticconcept = PROH \land normativestate\ norm = activated \land$
   $\quad norm.state \not\subseteq beliefs \land norm.deactivation \subseteq beliefs \Rightarrow$
   $\quad normativestate = normativestate \oplus \{(norm, deactivated)\} \land$
   $\quad normativesatisfaction = normativesatisfaction \oplus \{(norm, fulfilled)\}) \land$

5 : $(\,norm.deonticconcept = OBL \land normativestate\ norm = activated \land$
   $\quad norm.state \not\subseteq beliefs \land norm.deactivation \subseteq beliefs \Rightarrow$
   $\quad normativestate = normativestate\{(norm, deactivated)\} \land$
   $\quad normativesatisfaction = normativesatisfaction \oplus \{(norm, violated)\})$

In the context of the social system, if a new product is launched, *Norm 1* is activated (line 1). *SNA* promotes the product, *Norm 1* is fulfilled (line 2). *SNA* did not promote the product into 15 days after it has been launched, *Norm 1* is violated (line 5).

3.2 Norms Selector

The main goals of the *Norms Selector* function are to evaluate and select the activated norms that the agent intends to fulfill, and update the set of goals taking into account such norms. This function performs three tasks:

Detecting and Overcoming Conflicts: In the *DetectOvercomeConflicts* schema conflicts are detected in line 1. If two different activated norms, one being an obligation and the other a prohibition, specify the same state, they are in conflict. If the agent achieves the state, it will be fulfilling the obligation and violating the prohibition, but if it does not achieve the state, it will be violating the obligation and fulfilling the prohibition. The conflicts are overcome in lines 2 and 3 by choosing the norm that the agent is more motivated to fulfill. If the motivations have equal values, we can choose either one.

DetectOvercomeConflicts
ReviewDeActivatedNorms

$\forall n1 : norms \mid normativestate n1 = activated \bullet$
$\quad \exists n2 : norms \setminus \{n1\} \mid normativestate n2 = activated \wedge$
$1 : (n1.state = n2.state \wedge$
$\quad (n1.deonticconcept = OBLIGATION \wedge n2.deonticconcept = PROHIBITION) \vee$
$\quad (n2.deonticconcept = OBLIGATION \wedge n1.deonticconcept = PROHIBITION)) \bullet$
$2 : (deonticinfluence(n1.deonticconcept, n.state) +$
$\quad sanctionsmotivation(n1.rewards) + sanctionsmotivation(n2.punishments) \geq$
$\quad sanctionsmotivation(n2.rewards) + sanctionsmotivation(n1.punishments) \Rightarrow$
$\quad normativestate = normativestate \oplus \{(n1, tobe\_fulfilled)\} \wedge$
$\quad normativestate = normativestate \oplus \{(n2, tobe\_violated)\}) \wedge$
$3 : (\neg (deonticinfluence(n1.deonticconcept, n.state) +$
$\quad sanctionsmotivation(n1.rewards) + sanctionsmotivation(n2.punishments) \geq$
$\quad sanctionsmotivation(n2.rewards) + sanctionsmotivation(n1.punishments)) \Rightarrow$
$\quad normativestate = normativestate \oplus \{(n2, tobe\_fulfilled)\} \wedge$
$\quad normativestate = normativestate \oplus \{(n1, tobe\_violated)\})$

A conflict is overcome as follows. If the agent's motivation for fulfilling the first norm plus the motivation for violating the second norm is greater than or equal to the motivation for fulfilling the second norm plus the motivation for violating the first norm (see line 2), the first norm is selected to be fulfilled and the second one to be violated. When the opposite happens, the second norm is selected to be fulfilled and the first to be violated (see line 3).

The motivations for fulfilling or violating a norm is evaluated as follows. An agent evaluates its motivation for fulfilling a norm by summing the deontic (obligation or prohibition) influence on the state regulated by the norm (described in *deonticinfluence* function) and the motivation for receiving the rewards. The motivation for violating is basically the motivation for receiving the punishments. Both the motivations for fulfilling or violating a norm is evaluated by using the function *sanctionsmotivation*.

The deontic influence of a norm is evaluated by the following strategy: (i) if the norm is an obligation and the agent is motivated to achieve such goal (i.e., the motivation is positive), the influence of the norm is equal to the motivation of such goal. However, if the agent is not motivated to achieve such goal (i.e., the motivation is negative), the norm influences the agent negatively since it is obligating the agent to achieve a goal it does not want to; and, (ii) if the norm is a prohibition and the agent is not motivated to achieve such goal, the norm influences the agent positively. Otherwise, if the agent is motivated to achieve such goal, the norm influences the agent negatively since it disturbs the achievement of the agent's desire.

$\quad$ *deonticinfluence* : *DeonticConcept* × *State* → $\mathbb{Z}$

$\forall dc$: *DeonticConcept*; *state* : *State*; *inf* : $\mathbb{Z}$ •
1 : (*dc* = *OBL* ⇒ *deonticinfluence*(*dc*, *state*) = *goalmotivation state*) ∧
2 : (*dc* = *PROH* ⇒ *deonticinfluence*(*dc*, *state*) = −*goalmotivation state*)

At last, the rewards and punishments describe the achievement of environment states. So, the motivation for receiving the rewards or punishments is equal to the sum of the motivation of the agent's goals that specify the same states regulated by the rewards or punishments.

$\quad$ *sanctionsmotivation* : $\mathbb{P}$ *State*

$\forall states$: $\mathbb{P}$ *State*; *m* : $\mathbb{Z}$ •
sanctionsmotivation(*states*) = *m* ⇔ $\forall state$: *states* • *m* = *m* + *goalmotivation state*

Let's take Norm 2 and 3 and consider that both goals "to sell" and "to promote" of *SNA* have a motivation equal to "2", the goal "to increase reputation" has a motivation equal to "1", and the goals contrary to the ones that the agents want to bring about have motivation equal to "-1". **Norm 2** has one reward and one punishment and the agent's motivations associated with such norm is evaluated as follows: (i) given that Norm 2 is an obligation and the state regulated by it is equal to the goal "to sell" with motivation "+2", the influence of the norm is "+2"; (ii) the reward is the achievement of an environment state that is equal to the agent goal "to increase reputation" with motivation "+1", so such reward has contribution equal to "+1"; and (iii) the punishment is equal to the agent goal that has motivation equal to "-1", so such punishment has contribution equal to "-1".So the agent's motivation for fulfilling Norm 2 is "+3" and for violating it is "-1". **Norm 3** has one punishment and the agent's motivations associated with such norm is evaluated as follows: (i) given that Norm 3 is a prohibition and the state regulated by it is equal to the goal "to sell" with motivation "+2", the influence of the norm is "-2"; and (ii) the punishment is the achievement of an environment state that is equal to the agent goal with motivation "-1", so such punishment has contribution equal to "-1". So the motivation for fulfilling Norm 3 is "-2" and for violating it is "-1".

$\quad$ Finally, in order to better understand the overcome of conflicts, let's consider the situation where a new product is launched. The promoting rate of *SNA* related to any product is lower than 2 promotions per month. Norms 2 and 3 are activated and a conflict between them is detected and should be overcome. The conflict is overcome by selecting Norm 2 to be fulfilled and Norm 3 to be violated since the agent's motivation for fulfilling the first norm ("3") plus the agent's motivation for violating the second norm ("-1") is greater than agent's motivation for fulfilling the second norm ("-2") plus the agent's motivation for violating the first norm ("-1").

$\quad$ **Choosing Norms:** The *ChooseNorms* schema specifies the agent reasoning to decide whether to fulfill or violate the non-conflicting activated norms.

On one hand, the agent decides for fulfilling an obligation if the agent's motivation for achieving the state regulated by the norm plus the motivation for receiving the rewards is greater than the motivation for receiving the punishments. Otherwise, it decides for violating the obligation. On the other hand, the agent decides for fulfilling a prohibition if the agent's motivation for receiving the rewards minus the motivation for not achieving the state regulated by the norms is greater than the motivation for receiving the punishments. Otherwise, it decides for violating the prohibition.

ChooseNorm
DetectOvercomeConflicts

$\forall\, norm : norms \mid normativestate\,norm = activated \bullet \neg\ \exists\, n2 : norms \setminus \{n1\} \mid$
$\quad normativestate\,norm = activated \wedge DeonticConflict(norm, n1) \bullet$
$1 : (\ deonticn\!fluence(norm.deonticconcept, norm.state) +$
$\quad\quad sanctionsmotivation(norm.rewards) \geq$
$\quad\quad sanctionsmotivation(norm.punishments) \Rightarrow$
$\quad\quad normativestate = normativestate \oplus \{(norm, tobe\_fulfilled)\}) \wedge$
$2 : (\neg\ (\ deonticn\!fluence(norm.deonticconcept, norm.state) +$
$\quad\quad sanctionsmotivation(norm.rewards) \geq$
$\quad\quad sanctionsmotivation(norm.punishments)) \Rightarrow$
$\quad\quad normativestate = normativestate \oplus \{(norm, tobe\_violated)\})$

Let's consider that Norm 1 is activated and the motivation of the goals are as described before. The evaluated contribution for fulfilling Norm 1 is "3" and the contribution for violating is "-1". Thus, such norm is selected to be fulfilled.

Compliance with Norms: Once the agents have chosen the norms to be fulfilled and the ones to be violated, it is time to review the agent's goals (by generating new goals or updating their motivations), as described in Schema *ComplianceNorms*.

Making the agent aware of the norms rewards and punishments: If there is a goal equal to the state regulated by a set of norms (line 1), the motivation of such goal is updated by taking into account such norms as follows: (*Obligation* - see line 2) in case the agent chooses to achieve the obligation, the motivation of the goal is updated by summing the agent's motivation for receiving the rewards minus the motivation for receiving the punishments; (*Prohibition* - see line 3) in case the agent chooses to violate the prohibition, the motivation of the goal is updated by summing the agent's motivation for receiving the punishments minus the motivation for receiving the rewards;

Making the agent aware of the taken decisions: If there is not a goal equal to the state regulated by a norm (line 4), it's possible to distinguish two cases: (1) if the norm is an obligation selected to be fulfilled or a prohibition selected to be violated, a new goal equal to the state regulated by norm is generated with motivation defined as follows: (*Obligation Selected to be Fulfilled* - see line 5) if the norm is an obligation, the motivation of the state is updated by summing the agent's motivation for receiving the rewards minus the motivation for receiving the punishments; (*Prohibition Selected to be Violated* - see line 6) if the norm is a prohibition, the motivation of the state is updated by summing the agent's motivation for receiving the punishments minus the motivation for receiving the rewards; and (2) if the norm is an obligation selected to be violated or a prohibition selected to be fulfilled, nothing happens since if there is not a goal that is equal to the state regulated by norm, then it is not possible to fulfill the obligation or violate the prohibition.

ComplianceNorms
ChooseNorm

$1 : \forall\, norm : norms \bullet (\exists\, goal : goals \mid norm.state = goal \bullet$

$2 : (norm.deonticconcept = OBLIGATION \Rightarrow$

$goalmotivation = goalmotivation \oplus \{(goal, goalmotivation\ goal +$
$sanctionsmotivation\ norm.rewards - sanctionsmotivation\ norm.punishments)\}) \wedge$

$3 : (norm.deonticconcept = PROHIBITION \Rightarrow$

$goalmotivation = goalmotivation \oplus \{(goal, goalmotivation\ goal +$
$sanctionsmotivation\ norm.punishments - sanctionsmotivation\ norm.rewards)\}) \vee$

$4 : (\neg (\exists\, goal : goals \mid norm.state = goal \bullet$

$5 : (normativeselection\ norm = tobe\_fulfilled \wedge norm.deonticconcept = OBL) \Rightarrow$
$goalmotivation = goalmotivation \oplus \{(norm.state, sanctionsmotivation\ norm.rewards -$
$sanctionsmotivation\ norm.punishments)\} \wedge goals = goals \cup \{norm.state\}) \wedge$

$6 : (normativeselection\ norm = tobe\_violated \wedge norm.deonticconcept = PROH) \Rightarrow$
$goalmotivation = goalmotivation \oplus \{(norm.state, sanctionsmotivation\ norm.punishments -$
$sanctionsmotivation\ norm.rewards)\} \wedge goals = goals \cup \{norm.state\}))$

Let's remember that *SNA* has the goal "to promote" and the agent has selected Norm 1 to be fulfilled. The motivation of the "to promote" goal that is equal to the state regulated by Norm 1 is increased to 3. Such value is evaluated by summing the normative influence of Norm 1 (equal to "+2") and the contribution of its rewards (equal to "+1"):"2+1=3".

4 Discussion

This paper proposes an architecture model[2] to build motivation-oriented agents that can reason about norms. The model helps agents (i) on checking if they should adopt or not the norm; (ii) on checking the fulfillment or violation of a norm while verifying its activation and deactivation conditions; (iii) on evaluating the effects of the fulfillment or violation of the norm on their desires; (iv) on identifying and solving conflicts among norms; and (v) on complying with the norms by generating new goals or updating their priorities according to their choice of fulfilling or not a norm.

Our work was influenced by the agent model described in [3] and the architectural model proposed in [8]. In [3] the authors describe a *selfish* agent that decides for fulfilling or violating a norm by evaluating the positive and negative effects of rewards and punishments on their goals. The architectural model proposed in [8] presents functions to deal with the adoption of norms and the influence of norms on the selection of desires. However, such works are incomplete since they do not give support to the check of

---

[2] The architecture was implemented in Jason and is available at
http://wiki.les.inf.puc-rio.br/index.php/NormativeAgent

norms violated and fulfilled, and the detection and overcome of conflicts among norms, unlike our work.

The BOID (*Belief-Obligation-Intention-Desire*) architecture proposed in [2] is an extension to BDI that considers the influence of beliefs, obligations, intentions and desires on the generation of the agent's goals. The BOID architecture applies the notion of agent types to help with the generation of the goals.

Thus, this work adopts a *static approach* to deal with norms and goals based on the agent types. Our work uses the norm influence and the motivation of the desires to provide a quantifiable solution that enables the agent to deal with desires and norms according to the situation that it is exposed to.

The architecture for BO (Belief-Obligation) rational agents [5] follows a qualitative decision theory based on beliefs and obligations rules. The approach takes into account the beliefs and obligations while generating goals or decisions to achieve goals with the aim to maximize the set of goals that will be achieved. Such work does not consider prohibition norms, rewards and punishments but its results confirm the importance of a ANA agent to take into account the system norms while achieving their goals.

The authors in [4] and [6] present concepts used to model agents in environments governed by norms. Although such approaches consider that the selection of desires should be based on their priorities and they can be influenced by norms, the approaches do not present a complete strategy with a set of verification in the norm review process, and strategies to identify and overcome conflicts between norms, such as our work does.

References

1. Baekdal, T.: Social Commerce: It is a completely new playing field! Baekdal Plus (2011)
2. Broersen, J., Dastani, M., Hulstijn, J., van der Torre, L.: Goal generation in the boid architecture. Cognitive Science Quarterly (2002)
3. Castelfranchi, C., Dignum, F., Jonker, C.M., Treur, J.: Deliberative normative agents: Principles and architecture. In: 6th ATAL. Springer (2000)
4. Criado, N., Argente, E., Botti, V.: Normative Deliberation in Graded BDI Agents. In: Dix, J., Witteveen, C. (eds.) MATES 2010. LNCS, vol. 6251, pp. 52–63. Springer, Heidelberg (2010)
5. Dastani, M., Torre, L.V.D.: What is a normative goal? towards goalbased normative agent architectures. In: RAST. Springer (2004)
6. Dignum, F.: Autonomous agents and social norms. In: Proc. of the Workshop on Norms, Obligations and Conventions, pp. 56–71 (1996)
7. Diller, A.: Z: An Introduction to Formal Methods. John Wiley & Sons, Inc. (1990)
8. Lopez-Lopez, F., Marquez, A.: An architecture for autonomous normative agents. In: Proc. of the 5th Int. Conf. in Computer Science (2004)

A Coalition Formation Mechanism for Trust and Reputation-Aware Multi-Agent Systems

Bruno W.P. Hoelz and Célia Ghedini Ralha

University of Brasília, Brasília-DF, Brazil
bhoelz@unb.br, ghedini@cic.unb.br

Abstract. Most multi-agent systems engineering methodologies propose the clear definition of roles and organizations. However, in distributed environments where agents with distinct capabilities coexist and cooperate to solve problems, having a rigid organization structure makes the system less adaptable to changes and failures. Some of the approaches to deal with these difficulties include centralized coordination and planning and the use of homogeneous agent capabilities. These solutions oppose key benefits of multi-agent systems, especially the agents' autonomy to interact and organize freely. In this paper, a novel approach is proposed where agents form and dissolve coalitions in a service-oriented environment while maintaining their autonomy. This allows the agent society to adjust to the demand for services and react to failures. To achieve this flexibility, a coalition formation mechanism for trust and reputation-aware multi-agent systems is employed. As agents interact, they establish a network of trusted peers that allows them to form stable coalitions with reduced risk of failures. Agents can also expand this network by exploring new partnerships based on the reputation of unknown agents that are recommended by these known peers. Experiments were performed to evaluate the proposal, with positive results in environments up to fifty agents under varying service demand and failure rates.

Keywords: coalition formation, trust, reputation, multi-agent systems.

1 Introduction

Most multi-agent systems (MAS) engineering methodologies require the clear definition of roles and organizations during the design process. This results in a rigid structure that makes the system less adaptable to changes and failures. Some of the approaches to deal with this lack of flexibility include centralized planning and redundant agents. However, these solutions oppose key benefits of multi-agent systems, especially the autonomy of agents. More interesting solutions, such as dynamic coalition formation, seek to preserve this autonomy while providing the ability to respond to changes in the environment and adapt accordingly.

In this work, a novel approach is proposed where agents not only employ trust and reputation information to choose their partners, but they also seek to optimize these relationships by expanding their network of known agents and

L.N. Barros et al. (Eds.): SBIA 2012, LNAI 7589, pp. 162–171, 2012.

adjusting its notions of trust and reputation according to the results obtained. The agent's goal in the long term is to establish partnerships with reduce risk of failure and higher utility.

In this case, agents operate in a service-oriented environment where cooperation is needed to provide complex services that cannot be provided individually. In this environment, the MAS is constantly receiving service requests from multiple sources and must respond within established parameters of Quality of Service (QoS), such as timeliness and cost. Therefore, agents with distinct capabilities must form coalitions to perform these requests. Agents have the autonomy to choose their partners and to use their own trust and reputation models. In the absence of trust and reputation information, they can rely on a Coalition Formation Service (CFS) to introduce them to new agents. By using trust and reputation as a reflex of failures and changes in the demand for services, agents can adjust the size of coalitions and modify their partnerships to obtain better results.

The main goal of this work is to evaluate the impact of trust and reputation in the coalition formation process, and how agents can optimize trust in order to reduce the risks of cooperation in different scenarios. Experiments were performed to evaluate this, with positive results in environments with up to fifty heterogeneous agents, varying service demand and failure rates. Initially, in the absence of trust and reputation information, coalitions are formed mainly by using the CFS. Later on, agents choose to form coalitions based on their own models of trust and reputation. Thus, this dynamic approach to coalition formation, even in the absence of trust information, emphasizes flexibility over explicit definition of an organizational structure.

This paper is organized as follows: in Section 2, an overview of coalition formation is presented; Section 3 describes related works regarding coalition formation in MAS; the proposed approach is detailed in Section 4; Section 5 contains the experiments that were performed and an analysis of the results; finally, Section 6 presents the conclusions and future works.

2   Coalition Formation

In a MAS, agents can cooperate to solve problems that cannot be solved individually. Some problems may require a large number of agents and great effort is spent in coordination and communication. Several organizational paradigms are available, such as hierarchies, teams, and federations [1]. Each paradigm brings some benefits and drawbacks. Committing to a single organization may be simpler from a design standpoint, but it may take away some of its agents' autonomy. Whatever the choice of organization, additional interactions and protocols must be conceived to deal with known points of failure and weaknesses. These measures may restrict the agents' autonomy even further.

One organizational paradigm that can be employed to achieve more flexibility and adaptability is the coalition. Coalitions are dynamic, usually short-lived and formed with a specific purpose [1]. Every subset of agents represents a potential coalition that can be formed to solve a problem. Choosing a strategy for the formation and adaptation of coalitions can be complex, depending on the conditions imposed. Additionally, although coalitions with more members are capable of handling larger tasks, there is an increased cost in communication and coordination.

In the context of service-oriented computing, a multi-agent coalition can offer services that are combinations of services provided by its members. Additionally, Quality of Service (QoS) parameters can be introduced in the service request to better specify and evaluate the performed service. Consequently, agents in the coalition need to cooperate to increase their joint capacity to perform the requests within QoS conditions.

3   Related Works

In this section, some works related to coalition formation are introduced. These illustrate distinct approaches to the coalition formation process, each with its own characteristics and limitations.

A distributed approach to self-organization in a cooperative sensor network is presented in [2]. Agents form appropriate coalition through iterative negotiations. A one-level hierarchy is used, dividing the space in sectors with one manager each. Sector managers communicate among them using a single channel. Using a bottom-up coalition formation process, the system forms a set of disjoint coalitions. One important point that is cited is that each new member in a coalition represents an overhead, so adding agents indefinitely does not correspond to unlimited benefit.

A cooperative coalition formation algorithm guided by an underlying organization structure is presented in [3]. It employs a hierarchy for being both scalable and distributable. Agents are divided into managers and individuals. Once a manager receives a task, it must choose one of its children in the hierarchy to receive the task and it must inform its own manager of this allocation. If there are not enough resources to perform the task, the task is passed to the higher level manager, who can use its wider view of the organization to decompose and reallocate the task to other managers.

An algorithm for coalition formation for task allocation in both general multi-agent domains as well as multi-robot domains is presented in [4]. The algorithm expands on the work of [5], and considers two restricted scenarios: (i) coalitions that are restricted to a limited number of agents and (ii) coalitions that have identical agents of a specific number of types time. Both restrictions greatly reduce the complexity of the coalition formation problem. The first reduces the number of possible coalitions, while the second allows for the use of dynamic programming to find an optimal coalition structure in polynomial time.

4   Proposed Solution

To achieve the formation and adaptation of agent coalitions, a model with four main components is proposed: (i) a trust and reputation meta-model, (ii) a coalition formation service, (iii) a coalition formation algorithm using trust and reputation information, and (iv) an adaptation heuristic. The following sections present these components.

4.1   Trust and Reputation Meta-model

Autonomous agents often need to interact with other agents to achieve their goals. To choose its partners, an agent must have some knowledge of their capabilities and competence, either directly through previous interactions (experience-based trust) or by requesting information about their reputation from other agents (recommendation-based trust). Several models of trust and reputation have been proposed, and extensive reviews can be found on the works of [6] and [7]. According to [8], trust can be used to assess risks in cooperation, representing an estimate of the likelihood of failure. Regardless of the trust model used, any episode of misplaced trust must have an impact in the beliefs of the trusting agent in order to allow it to adapt and improve its partnerships. Consequently, with the increase in trusted interactions, the expected risks and likelihood of failures should be lower.

An important property of a trust model, as noted by [9], is that each agent should be able to evaluate trust by itself. We take that as a premise, since in a MAS with heterogeneous agents, such diversity is only natural and regardless of the model used by an individual agent, it must be able to interact and share information. If it fails to do so, its own reputation will be affected. Finally, this diversity gives the MAS designer the opportunity to observe, as the agent society progresses, the practical differences in the models used and how they interact.

Thus, in our proposal we define a minimal meta-model of trust and reputation to allow agents to use their own models, while still being capable of interacting to form coalitions. The meta-model supports both experience and recommendation-based trust. The trust T that agent A has on agent B is calculated as a weighted product presented in Equation 1.

$$T(A, B) = \frac{w_e \times T_e(A, B) + w_r \times T_r(A, B)}{w_e + w_r} \qquad (1)$$

where w_e and T_e are respectively the weight of experience-based trust and its value, and w_r and T_r are related to recommendation-based trust. Agents can use an arbitrary scale for the values of T. However, values are normalized to real numbers in the interval of $[0, 1]$ for sharing.

Agents can adopt a multi-dimensional model for trust [8], by maintaining separate values for each Quality of Service (QoS) parameter used in the system. However, since other agents may not use the same approach, when querying for specific dimensions, the queried agent may return the overall reputation value. To effectively share specific values, agents must share the same concepts of dimension, usually by means of a shared ontology. The average trust value obtained from multi-dimensional trust (MDT) is given by Equation 2.

$$MDT(A, B) = \sum_{i=1}^{|Q|} w_{q_i} \times T_{q_i}(A, B) \qquad (2)$$

where $q_i \in Q$ is a trust dimension corresponding to a QoS parameter and w_{q_i} is the weight of this parameter. Again, values must be normalized for sharing.

Additionally, the meta-model defines four policies that a trust model employed by an agent must have:

- basic predisposition to trust: this basic trust level varies from agent to agent and is adjusted according to environmental conditions and interactions over time. New agents are trusted considering the level of basic trust as a vote of confidence. Otherwise, they would not have a chance to interact and establish direct trust;
- trust maintenance: agents must define how they update trust in the case of success or failure. When the agent is part of an organization, it can update trust differently for each member. This can be done if the utility of a partner agent is perceived as higher than the others. Likewise, in the case of a failure, an agent may be blamed for the failure and have its trust reduced by a higher amount;
- temporal decay of trust: both trust and reputation must not be static. They must be renewed with time to reflect the current conditions. To achieve that, an agent's trust policy must include a mechanism to update trust with the passage of time.
- reputation verification: reputation information received from others must be verified. Since agents may have incentives to provide distorted reputation information, agents must have a mechanism to check the received information.

4.2 Coalition Formation Service

To bootstrap coalition formation in the absence of trust information, agents can make use of a coalition formation service (CFS) offered by the MAS. Agents that have no knowledge of other agents' reputations, such as newcomers, or agents whose trusted partners are currently busy can use this service to explore new partnerships. With the formed coalition, members negotiate a joint proposal. If the coalition is awarded the contract, it performs the service and afterwards the agents can update the trust in their partners according to their own trust maintenance policies.

To use the service, agents subscribe to topics in the CFS corresponding to the services they provide. The CFS then uses a greedy algorithm, shown in Listing 1.1, to quickly match agents to service requests. The algorithm tries to form coalitions with agents that cannot perform the request alone. In the end, the candidate solutions are returned and the one is chosen according to the criteria determined by the client (e.g. lowest cost within a time limit).

Listing 1.1. The Coalition Formation Service algorithm

```
1   Algorithm CFS (request, subscribers):
2       while (subscribers not empty):
3           pick a random agent from the subscribers list;
4           if the service can be performed by the agent
5               within the QoS parameters:
6               add agent as a candidate solution;
7               remove agent from subscribers;
8           else:
9               add agent to a temporary coalition;
10              remove agent from subscribers;
11              if the request is fully allocated to the temporary coalition:
12                  add temporary coalition as a candidate solution;
13                  start new temporary coalition;
```

4.3   Coalition Formation with Trust and Reputation Information

Agents that have trusted partners can negotiate the formation of a coalition directly with them. To do this, agents use the coalition formation with trust and reputation information (CFwTRI) algorithm shown in Listing 1.2.

Listing 1.2. The CFwTRI algorithm

```
1   Algorithm CFwTRI:
2       choose a trusted or recommended agent;
3       invite agent to join/form coalition;
4       if agent doesn't accept the invitation and I'm left alone:
5           subscribe to CFS;
6       else if agent is already in a coalition:
7           if agent's coalition is approved:
8               form/join coalition;
```

The CFwTRI algorithm allows the formation of coalitions in a trusted context. Each decision in the algorithm is guided by an stochastic process with varying probability, detailed in the next section. These probabilities depend on the agent's perception of trust. The expansion of a coalition is only possible if the level of trust increases, otherwise agents will not perceive any advantage in partnering with known agents instead of relying on the CFS.

Before choosing which agent to invite, agent A can request a recommendation from one or more trusted agents. Agent A can combine reputation information from these sources to decide whether to risk a partnership with the unknown but recommended agent B. When proposing a partnership with B, agent A informs B of its sources. This allows the other agent to verify the reputation of A using the same sources. By exploring partnerships with recommended agents, the agents expand their network of known, trusted (or at least recommended) agents.

Each coalition lives for at least one service request. If a contract is not awarded to the coalition or after the service is performed, its members can negotiate the maintenance of the coalition for the next request or simply disband. Even the intention to explore new partnerships and coalitions depends on trust. The higher the trust in the members of a coalition, the less likely an agent is to leave it, since the perceived risk is lower. With time, some coalitions will grow stable and stay unchanged for several requests.

For example, if an agent A invites another agent B, and both are currently in a coalition, the invitation must be accepted not only by A and B, but also by the members of both coalitions. This means that the chances of merging two large coalitions are very low, except if trust among the members of both coalitions is very high.

4.4   Adaptation Heuristic

To obtain a flexible, adaptive coalition formation mechanism, agents must react to the results of its interactions and to changes in the environment, such as the arrival of new agents, changes in the demand for services or failures. This process depends on a series of decisions made by the agent considering probabilities that take into account the trust it has on other agents, either through direct interaction or reputation.

To guide the agent's decisions, a heuristic based on simulated annealing [10] is proposed. In this heuristic, an agent A decides whether to form a coalition or resort to the CFS depending on p, given by Equation 3:

$$p = exp\left(\frac{-\Delta E}{\tau_i}\right) \tag{3}$$

where ΔE is the energy parameter of the simulated annealing process and is calculated by the variation of the average trust of known agents between two iterations $E_i - E_{i-1}$. When trust decreases, $p = 1$, then the agent uses the CFS. If trust increases, the probability p is used to choose a known agent or a recommended agent (with probability $1 - p$). Since low energy states have higher probability of occurrence, as trust increases the use of the CFS will happen less often.

In Equation 3, τ is called the temperature parameter and is used to control the rate of convergence of p. It is updated after each iteration i (that represents the performing of a request), such that $\tau_i = \alpha\tau_{i-1}$, where α is the temperature adjustment factor. Values of τ and α will vary among agents. Higher values of τ and lower values of α give agents more time to explore new partnerships and establish a larger trust network. A very low value of τ or high value of α, on the other hand, pushes the agent faster towards previously known partners.

A known agent B is chosen by agent A with probability $\rho(A, B)$ equal to its relative trust calculated by Equation 4.

$$\rho(A, B) = \frac{T(A, B)}{\sum_{i=1}^{|K|} T(A, k_i)} \tag{4}$$

where $k_i \in K$ is an agent known to A. Consequently, partnerships with well trusted agents are more likely to repeat. If the agent seeks a recommendation, the referrer R is chosen from the pool of trusted agents using the same probability $\rho(A, R)$. The referrer again recommends an agent S from its own pool of trusted agents again with probability $\rho(R, S)$. So the probability of an unknown agent S being recommended to A is given by $\delta(A, R, S)$ in Equation 5.

$$\delta(A, R, S) = (1 - p) \times \rho(A, R) \times \rho(R, S) \tag{5}$$

In the first case, the invited agent B accepts the invitation from A with probability $\rho(B, A)$. In the second, the invited agent S accepts the invitation from A through R with probability $\delta(S, R, A)$.

After the service is performed by the coalition, the agent can choose to leave the coalition with probability q that is calculated similarly to p by using Equation 6.

$$q = exp\left(\frac{\Delta E - \Delta T_G}{\tau_i}\right) \tag{6}$$

where ΔE and ΔT_G are respectively, the moving averages of basic trust and the average trust in the coalition members in a given number of iterations established by the agent. This number of iterations impact how agents perceive changes in the coalition's trust. The higher the trust in the organization relative to the basic trust (that includes all other known agents), the less risky the coalition is and the less likely is an agent to leave.

5   Experiments and Results

In order to test the proposed approach, we built a prototype of a service-oriented MAS to perform the experiments. In this context each agent A is defined as $A = \{S_A, t_b, T_d, T_r, c, \tau_0, \alpha\}$, where S_A is the service offered by A, t_b is the basic trust level, T_d is the set of trust values obtained through direct interaction, T_r is the set of trust values obtained from reputation information and c is the agent's current coalition. The initial temperature of the simulated annealing process is given by τ_0, while α is the temperature adjustment factor.

Fifty agents were used in the tests. The services they provide are chosen randomly from a group of three distinct services. Their capacity to perform the services, t_b, τ_0, and α are set randomly from normal distributions in order to diversify the agents' use of the proposed mechanisms. In all experiments, service requests with varying sizes are generated and published in the system. If a coalition or agent does not deliver the results of a contracted service within the specified time limit, this is considered a failure and the value for the service is not awarded.

In the first experiment, we test the interactions of agents from the start of the MAS operation, in the absence of trust and reputation information. A second experiment was performed to observe the evolution of the average trust in the system over time. Figure 1 presents the results of the use of the coalition formation mechanisms and the average trust over time. It can be observed that agents initially rely on the coalition formation service, since no other agents are trusted. As the request progress, the other agents start to seek trusted or recommended agents to partner with and the coalition formation service stops being used. Only in the event of frequent failures the service is used again, since the overall trust across the system becomes lower. Agents or coalitions that can provide the service by themselves, do not rely on any of these mechanisms.

In the second experiment, four scenarios were established: small and large service requests with lower failures rates (SS, LF and LS, LF), and small and large service requests with high failure rates (SS, HF and LS, HF). A small service request, in this case, is a request that requires at most 4 agents to be performed. Failure rates are considered low when they are below 1%. In the case

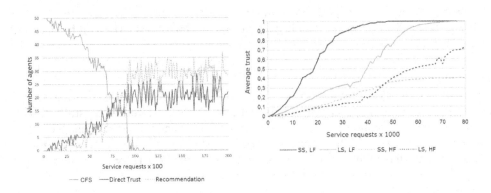

Fig. 1. Use of coalition formation mechanisms and average trust over time

of higher failure rates, it can be observed that trust still increases, albeit more slowly. This is a result of the coalition adaptation heuristic, as agents establish increasingly trusted and consequently more successful partnerships. With lower failure rates, both in the case of small and large requests, the average trust almost reached complete trust. In the case of large requests, this happens more slowly than with small services. Since small services depend on fewer agents, the impact of a failure on trust is lower.

6   Conclusions

In this paper we presented a novel approach to coalition formation in a trust and reputation-aware MAS operating in a service-oriented environment. Agents maintain information about trust and reputation of other agents in order to choose the best partners to perform complex services. As in the well known exploration-exploitation dilemma, agents weigh the risks and benefits of seeking new partnerships or repeating known ones. As time progresses, the coalitions grow and a higher level of trust is established among its members. As a result, the overall rate of failure is reduced. This allows the coalitions to react to changes in the demand for more complex services. If frequent failures occur in a coalition, agents can leave it to try new, more successful partnerships.

The results show that trust and reputation can be used to obtain flexible, adaptable coalitions that respond to changes in the environment. By forming coalitions in a trusted context, the chance of delivering results within the QoS parameters defined in the service requests increase. Results show that even with a higher frequency of failures, agents reorganize and the average trust keeps increasing with the successful performance of services, especially with services that can be performed by smaller coalitions.

The current work presents a mechanism that is both centralized (using the CFS) and distributed (using the CFwTRI algorithm). Differently from the works of [5] and [4], there is no need to establish restrictions in the number of member in the coalition or types of agents in the system. The use of trust and reputation in the agent's decision process reduces the dimension of candidate coalitions, by restricting partnerships to known (or at least recommended) peers. This can be decisive in the matter of scaling the system for hundreds or thousands of agents, and to test the applicability in large-scale distributed applications is a subject of future works.

Additionally, this work avoids assumptions of a stable environment that does not encompass the dynamics of a distributed environment, such as fluctuations in the cost of communication and the occurrence of failures, that can affect the utility of coalitions. By having trust and reputation values reflect the QoS, these factors are accounted for, even if indirectly.

Future studies includes a further experimentation to evaluate the optimality and stability of obtained coalitions, as well as a comparison to other coalition formation mechanisms that tackle real world scenarios.

References

1. Horling, B., Lesser, V.: A survey of multi-agent organizational paradigms. The Knowledge Engineering Review 19(04) (2005)
2. Sims, M., Goldman, C.V., Lesser, V.: Self-organization through bottom-up coalition formation. In: Proceedings of the Second International Joint Conference on Autonomous Agents and Multiagent Systems (2003)
3. Abdallah, S., Lesser, V.: Organization-based cooperative coalition formation. In: Proceedings of the IEEE/WIC/ACM International Conference on Intelligent Agent Technology, pp. 162–168 (2004)
4. Service, T., Adams, J.: Coalition formation for task allocation: theory and algorithms. Autonomous Agents and Multi-Agent Systems 22, 225–248 (2011)
5. Shehory, O., Kraus, S.: Methods for task allocation via agent coalition formation. Artificial Intelligence 101(1-2), 165–200 (1998)
6. Keung, S.N.L.C., Griffiths, N.: Trust and reputation. In: Griffiths, N., Chao, K.M. (eds.) Agent-Based Service-Oriented Computing, pp. 189–224. Springer (2010)
7. Sabater, J., Sierra, C.: Review on Computational Trust and Reputation Models. Artificial Intelligence Review 24(1), 33–60 (2005)
8. Griffiths, N.: Task delegation using experience-based multi-dimensional trust. In: Proceedings of the Fourth International Joint Conference on Autonomous Agents and Multiagent Systems, pp. 489–496 (2005)
9. Huynh, T.D., Jennings, N.R., Shadbolt, N.R.: An integrated trust and reputation model for open multi-agent systems. Autonomous Agents and Multi-Agent Systems 13(2), 119–154 (2006)
10. Kirkpatrick, S., Gelatt, C.D., Vecchi, M.P.: Optimization by simulated annealing. Science 220, 671–680 (1983)

Profile Recommendation in Communities of Practice Based on Multiagent Systems

João Luis Tavares da Silva, Sidinei D. Lubenow, and Alexandre M. Ribeiro

Universidade de Caxias do Sul (UCS), Caxias do Sul, RS, Brazil
{jltsilva,Alexandre.Ribeiro,eboff}@ucs.br

Abstract. This work describes the study, analysis, modeling and implementation of information agents aiming to manage and share knowledge in a context of virtual Communities of Practice. The main goal is to define a decentralized platform to manage profiles of users and communities. Two types of agents have been modeled, one including the behaviors needed to agents representing communities and other including the behaviors needed to agents representing members of communities. These behaviors have access to the profiles, what allow agents to exchange information about interest and knowledge in order to receive recommendations according to similarities found. Some experiments are presented showing how the system works in a specific domain and the results obtained.

Keywords: Communities of Practice (CoP), Recommendation, Multiagent Systems.

1 Introduction

Nowadays, the notion of community is a major focus of attention in Social Sciences, Business, Administration, Economics, Political Science and Free Software development [7,6,12]. The organizations tend to explore the use of collaborative technologies to develop new organizational capabilities, because the increasing development of technology tools for corporate portals and electronic communication. According to Wenger [15], one of the most efficient ways to develop these skills is through organizational Knowledge Management based on collaborative work through *Communities of Practice (CoP)*.

The main problem that we focus in this article concerning the amount of knowledge available into these kind of communities, which can cause an overload of information in the access of relevant information to social learning. Search engines are commonly used in these cases and recommendation systems are responsible for providing data automatically, in order to deal with the problem of information overload [13]. However, traditional works about recommendation systems (content or preferences recommendations) use static parameters that depend on external factors such as initial data or evaluations of others, being also ineffective in dynamic and distributed environments [9]. Much of the traditional techniques of recommendation gives privilege to individuals, relegating group

L.N. Barros et al. (Eds.): SBIA 2012, LNAI 7589, pp. 172–181, 2012.

recommendations to very specific works [5]. Also in [9] approaches using multi-agent systems are alternatives to group recommendation outlining the problems of information distribution.

This work is part of a research project of a conceptual framework for building and managing communities of practice [14]. The main goal of this framework is to provide a specification and a platform for creating and managing CoP, providing resources for a recommendation and management of knowledge through colla-boration tools and the interactions among its members. User and community profiles are represented by a proposed reference ontology. The definition of these profiles aims to provide all information identifying the main structural features of the participants in a particular CoP. The main purpose of these requirements is to define an approach to manage CoP and to find similarities between the profiles (interests/preferences), which supports the recommendation of partners and communities in a particular field of interest. The main factor considered is the management of static and dynamic profiles. The static profile is informed by the user while the dynamic profile is inferred from interactions with other members and communities.

This article takes into account two key issues in recommendations in collabo-rative learning communities: a unified representation of users and communities and the need to deal with heterogeneous and distributed sources of knowledge. The first issue uses the CoP Framework to explore the ontological representation of the members and community profiles. The second, uses an agent-based ap-proach for developing this CoP framework, so that the user's profiles of interest (preferences) and community profile are managed by virtual agents (personal assistants), where each virtual agent represents the user's interests.

2 Profile Management System

In this approach, each user is represented by an agent that manages the profile in terms of beliefs and knowledge that it has, whether individually or according to the communities in which the user participates. Thus, the agent can communi-cate and negotiate with other agents based on its beliefs and suggest relationships that have a common level of similarity through the interactions among agents.

The work in [10] specifies the goals, scenarios, features, perceptions and roles for the system using Prometheus methodology [11] for specifying and modeling agents and their functionality. Many of the agent-based methodologies covers phases from specification to architectural and detailed designs but through dif-ferent aspects of the agent development. Some of them often depicted all the phases on the same schema, difficulting readings (TROPOS [2]), whilst oth-ers are too general having adopted UML diagrams (such as ADELFE [1] or MESSAGE [3]). Prometheus offers supporting tools that verifies features at the architectural and detailed design level and it is mainly concerned with interfaces compatibility and plans consistency. From an AI planning-execution perspective, Prometheus provides an especially rich goal-driven approach for BDI-like agents.

Two types of agents were modeled: an agent representing the user of a com-munity and another agent representing the community itself. In this context, an

agent acts as a community manager of the dynamic profile of the group of participants and as an aggregator of social knowledge of the group. Figure 1 illustrates
the relationship between the two types of agents, besides the capabilities, and
also, the functionality required, according to the Prometheus methodology.

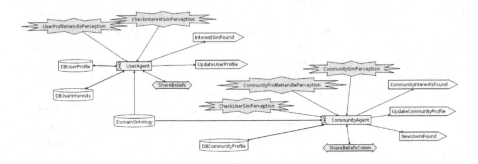

Fig. 1. Overview of the relationship between the User and User Community agents
with their capabilities, perceptions and interactions

The main features are:

- *User profile management*: is the validation of user information in order to
  store them on the agent belief base;
- *Agent recommendation*: occur through the message exchanges between User
  agents in order to find knowledge similarities based on common interests;
- *Community profile management*: is the management of the local community
  which is made by interest and a summary of the user's profile connected to
  it;
- *Search for potential users* to a community that are not yet part of it through
  an intersection/similarity of interests.

Communication between the environment and the agents is made through the
perceptions, which are the events that an agent can receive, diagnose and respond. Figure 1 also illustrates the perceptions of each type of agent. The perceptions of the *U*ser agent allow it to keep profile data belonging to a user, using
the *UserProfileHandlePerception*, and send messages to other agents who belong
to the same communities in order to seek similarities of knowledge through the
CheckInterestSimPerception. The community agent represents the administrator
of the community, and it has the following perceptions: *CommunityProfileHan-
dlePerception* to maintain its profile, *CommunitySimPerception*, to find similarities between communities, and *CheckUserSimPerception*, which searches for
users with the potential to develop the community. The features have goals that
must be performed, where each goal has its respective subgoals.

2.1 $\mathcal{U}$ser Profile Management

The $\mathcal{U}$ser agent is responsible for maintain the user profile and provide recommendations of other agents with related profiles. This agent has as core capabilities *User Data Maintenance* and *Intersection between users*.

The Algorithm 1 aims to return a list of potential users with information related to a source agent ($Agent_O$) to perform the *CheckInterestSimPerception* capability. This similarity relation of knowledge is verified by the plan *IntersectionUserPlan* through all users who participate in the same communities of the $Agent_O$. This request can be performed by the user or automatically by its agent in cycles of pre-determined periods of time.

The algorithm has an initial set of concepts from the ontological user/community profile ($SourceAgentProfile$) where each concept represents an interest of the agent, to which is assigned an initial activation value. This interest value is updated each time that agents show interests by the concept through any interaction in the community. Interactions occur through the use of the Web tools in the community, such as forums, sharing content, content postings, comments in the blogs, and other community contents.

The $AG(Interest_i)$ represents the list of $Agent_O$ that contains the j users with similar interests. The $Interest_i$ contains the term representing the interest of the $Agent_O$ profile. The $CheckOntologicalBase(t)$ process performs the analysis of the term (t) on the ontology base. The process $sendUserAgent_j(Interest_i)$ sends to the j agent the information returned by the ontology so that the others agents can check in their belief base by the compatibility with their own interests. The $SourceCommunity$ represents all communities in which the $Agent_O$ participates. The $receiveUserAgents()$ process receives the related interests from the j agents. The $Score_j$ stores the j agent's degree of interest on the term $Interest_i$.

$CalculateInterestScore()$ is computed by the Algorithm 2 that calculates the degree of interest with the j agent, taking into account only the dynamic information, i.e. knowledge created by the user/$Agent_O$ to participate in the community. This algorithm is used for both $\mathcal{U}$ser and $\mathcal{C}$ommunity agents in order to calculate the recommendation score based on the interests in question.

2.2 $\mathcal{C}$ommunity Profile Management

The $\mathcal{C}$ommunity agent has to search for recommendations from other communities, making them available to the administrator in order to manage profile information and find new users who matching the community profile. The perceptions, actions and databases related to the capabilities are also illustrated in Figure 1. The $\mathcal{C}$ommunity agent capabilities are: *Community Maintainability*, *Intersection Between Communities* and *Ability to Find New Users*.

Algorithm 3 aims to return a list of possible communities with information related to $Agent_O$. The profile similarity is calculated through all communities. This request can be made by the administrator or automatically by the community agent at predefined periods of time.

Also in this plan, the *CalculateCommunInterestScore()* process is triggered by the algorithm 2 with the difference that are the data of community agents that are considered in the recommendation score, based on the interests in question.

The algorithm 4 performs an intersection search for interests from the users who belong to the same community group that it is requesting. The matching profiles with the community are recommended.

In this plan, the *CalculateInterestScoreNewUsers()* process uses the algorithm 2 to calculate the similarity of interest with the user agents from all communities who *Agent$_O$* knows.

3   Application Scenarios

In the section we present the main scenarios illustrating the recommendations of users to users, communities to communities and recommendations of new users to communities. The profile of users and communities are represented in XML format. In the development phase was used the *Prometheus Design Tool* (PDT)[1] in order to generate the initial Java code skeleton, originally oriented for JACK platform. It was later made the portability to the JADE[2] platform, with which the prototype was developed.

In the scenarios involved, the agent's beliefs represent the user's interests and the interests of the community, which are private. In this case, users and communities have no access to the entire profile of the others in order to preserve user privacy. The only information available is a list of users or communities that can be recommend to the source agent who trigger the request.

In this paper the scope of representation of the ontology is limited to the user/community interests. We have distributed fifteen user agents in six community agents. For sake of space we do not describe all the agents but we give some examples. User and Community Agents have interests on the field of Programming Languages. For instance, Figure 2 represents the "static" Profile of some user and community agents.

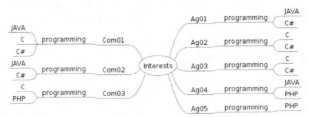

Fig. 2. Representation of the scope of the user agents and community profile with interests in a specific domain

The user agents are randomly distributed in the communities. For instance, the associated user agents to *Com01* community agent is *Com01* = {*Ag01*,

---

[1] http://www.cs.rmit.edu.au/agents/pdt/

[2] http://jade.tilab.com/

$Ag02$, $Ag03$, $Ag04$, $Ag06$, $Ag13$} while in $Com04$ agent is $Com04 = \{Ag03,$ $Ag13$, $Ag15$}. User agents may participate in several communities. The $Com01$ community agent has interest in "Programming", but specifically on $JAVA$, $C\#$ and C languages, while the preferences and profiles of some of its agents are illustrated in Figure 3, represented as a mind map.

Figure 3 illustrates a representation of the beliefs of each agent on the domain. For example, the agent $Ag01$ has specific knowledge about Java, C, C# and PHP languages. In each of the languages, the agent manifest an interest about some terms of the domain, through interaction tools on the portal. For example, in Figure 3 the $Ag01$ keeps some interests or doubts about the use of *Buttons* in Java and C#. Numbers in parentheses represent the frequency of use of this terms (activation values) obtained from the interaction tools.

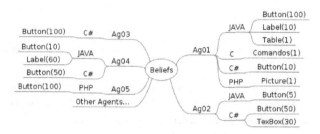

Fig. 3. Representation of the beliefs of the agents obtained from the interaction tools

Scenario 1: User Recommendation: This scenario shows the recommendation for user agents. The $Ag01$ agent represents the agent source, who searches for other similar agents in the community $Com01$. When an agent participates in several communities, all agents of the related communities are also considered. The agents who are able to receive and reply to this message are the $Ag02$, $Ag03$, $Ag04$ and $Ag13$ agents. Figure 4 illustrates the calculated results for the similarity of interest on their profiles.

The level of interest is obtained by Algorithm 1. In this scenario, the belief of the agents in $JAVA(Button)$ is $Ag02(5)$, $Ag04(10)$ and $Ag13(45)$. The $Ag13$ is the first agent in the recommendation list for $Ag01$. In this algorithm, $OntologyWeight$ has the value 1, without altering the importance of the belief. In other cases, the $OntologyWeight$ acts as a balance factor of the concepts of the domain in terms of ontologies. In fact, the weight of the terms in the ontology can be parameterized according to the importance of the community domain.

Scenario 2: Community Recommendation: This scenario describes the simulation of recommendation from communities where the community $Com01$ trigger the community recommendation plan. All the Community Agents interact to each other using the interests in their profile.

Figure 5 illustrates the result of the agents who present similarity of interests. In this scenario, the $Com02$, $Com03$, $Com04$, $Com05$, and $Com06$ agents replied the request for the terms $C\#$, Java, C. For instance, the $Com02$ and $Com06$

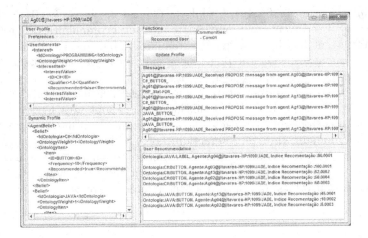

Fig. 4. Simulation of the recommendation of an user to a community

agents found interest similarity concerning the term *JAVA programming*. The results in the "Community Recommendation" are sorted by the recommendation score computed by Algorithm 3 using *OntologyWeight* value to classify the weight of interest from each community.

Fig. 5. Results of the recommendation of communities to communities

Scenario 3: Recommendation of New Potential Users: This scenario presents how the *C*ommunity agent may find out new users for their communities. In this simulation, the *Com*06 community agent sends messages for all agents participating in all the others communities with the same interests terms. The agents calculate the similarity of interest in those terms presented in their profile ontology. Figure 6 shows that the *Com*06 community has interest in *JAVA* programming, so the agents who has the same interest in their profiles are the *Ag*14, *Ag*01, *Ag*15, *Ag*13, *Ag*04 and *Ag*10 agents for *JAVA* programming). All those user agents are potential candidates to integrate *Com*06 community.

Fig. 6. Results of the simulation on the potential new agents for the *Com*03 community

4   Conclusion and Perspectives

This paper presented the specification, modeling and development of a multia-gent system for knowledge sharing in communities of practice. The main focus on the work are on managing user profiles from a website containing users and communities and the recommendation to users based on the similarity degree of interests into the profiles.

Most of the agent-based solutions needs some kind of collector agents [8,4], which running at background in order to feed a central recommendation base. In this paper the recommendations are distributed over the agents based on their recent interactions. Other issue in based-agent approaches to recommendation is that the agent would need a lot of knowledge from user to answer the user requests. Our ontological profile representation intends to avoid handling much information or a huge set of keywords from user.

From this model, it is possible to make recommendations based on the knowl-edge that each agent has, by mapping it into a structure indexed by an ontology. Thus, one can standardize the profile of interest by the categorization of interests addressed by a domain ontology.

In this paper, the criteria defined for quantify the similar terms between agents are maximized by a weight node in the ontology to which the term belongs. In the future work, we intend to extend the recommendation capabilities using an OWL[3] format representation for the profile ontology.

---

[3] http://www.w3.org/TR/owl-features/

References

1. Bernon, C., Gleizes, M.-P., Peyruqueou, S., Picard, G.: Adelfe: a methodology for adaptive multi-agent systems engineering. In: Petta, P., Tolksdorf, R., Zambonelli, F. (eds.) ESAW 2002. LNCS (LNAI), vol. 2577, pp. 156–169. Springer, Heidelberg (2003)
2. Bresciani, P., Perini, A., Giorgini, P., Giunchiglia, F., Mylopoulos, J.: Tropos: An agent-oriented software development methodology. Autonomous Agents and Multi-Agent Systems 8, 203–236 (2004)
3. Caire, G., Coulier, W., Garijo, F.J., Gomez, J., Pavón, J., Leal, F., Chainho, P., Kearney, P.E., Stark, J., Evans, R., Massonet, P.: Agent Oriented Analysis Using Message/UML. In: Wooldridge, M.J., Weiß, G., Ciancarini, P. (eds.) AOSE 2001. LNCS, vol. 2222, pp. 119–135. Springer, Heidelberg (2002)
4. Huang, L., Dai, L., Wei, Y., Huang, M.: A personalized recommendation system based on multi-agent. In: Proceedings of the 2008 Second International Conference on Genetic and Evolutionary Computing, WGEC 2008, pp. 223–226. IEEE Computer Society, Washington, DC (2008)
5. Jameson, A., Smyth, B.: Recommendation to Groups. In: Brusilovsky, P., Kobsa, A., Nejdl, W. (eds.) Adaptive Web 2007. LNCS, vol. 4321, pp. 596–627. Springer, Heidelberg (2007)
6. Lave, J., Wenger, E.: Situated learning: Legitimate peripheral participation. Learning in Doing 95(2), 138 (1991)
7. Lerner, J., Tirole, J.: The open source movement: Key research questions. European Economic Review 45(4-6), 819–826 (2001)
8. Lesser, V., Horling, B., Klassner, F., Raja, A., Wagner, T., Zhang, S.: BIG: An Agent for Resource-Bounded Information Gathering and Decision Making. Artificial Intelligence Journal, Special Issue on Internet Information Agents 118(1-2), 197–244 (2000)
9. Lorenzi, F., Bazzan, A.L., Abel, M., Ricci, F.: Improving recommendations through an assumption-based multiagent approach: An application in the tourism domain. Expert Systems with Applications 38(12), 14703–14714 (2011)
10. Lubenow, S.D.: Gestão de perfis em Comunidades de prática em Sistemas Multiagentes. Trabalho de conclusão (curso de ciência da computação), Universidade de Caxias do Sul, Caxias do Sul (2010)
11. Padgham, L., Winikoff, M.: Developing Intelligent Agent Systems. John Wiley & Sons, Melbourne (2004)
12. Preece, J., Maloney-Krichmar, D.: Online communities: Design, theory, and practice. Journal of Computer-Mediated Communication 10(4) (2005)
13. Resnick, P., Iacovou, N., Suchak, M., Bergstrom, P., Riedl, J.: Grouplens: an open architecture for collaborative filtering of netnews. In: Proceedings of the 1994 ACM conference on Computer Supported Cooperative Work, CSCW 1994, pp. 175–186. ACM, New York (1994)
14. da Silva, J.L.T., Moreto Ribeiro, A., Boff, E., Primo, T., Viccari, R.M.: A Reference Ontology for Profile Representation in Communities of Practice. In: García-Barriocanal, E., Cebeci, Z., Okur, M.C., Öztürk, A. (eds.) MTSR 2011. CCIS, vol. 240, pp. 68–79. Springer, Heidelberg (2011)
15. Wenger, E., Mcdermott, R., Snyder, W.: Cultivating communities of practice: A guide to managing knowledge. Harvard Business School Press (2002)

Algorithm 1: Calculate Generic Interest Similarity

Input: $SourceAgentProfile$
Output: List of recommended agents for $Agent_O$.
begin
    foreach $t \in SourceAgentProfile$ **do**
        $Interest_i \leftarrow CheckOntologicalBase(t)$;
        foreach $Agent_j \in SourceCommunity$ **do**
            $sendUserAgent_j(Interest_i)$;

        foreach $Agent_j \in SourceCommunity$ **do**
            $Interest_j \leftarrow receiveUserAgents()$;
            $Score_j \leftarrow CalculateInterestScore()$;
            $AG(Interest_i) \leftarrow AG(Interest_i) \cup AG(Interest_j) + Score_j$;

    $Return \leftarrow AG(Interest)$;

Algorithm 2: Calculate the profile intersection among agents

Input: $SourceAgProfile$ and $TargetAgProfile$.
Output: $Score$ with a recommendation index.
begin
    $Score \leftarrow 0$;
    foreach $interest \in SourceAgProfile \cap TargetAgProfile$ **do**
        while $interest \in OntologyBase$ **do**
            $Score \leftarrow \sum_{i=1}^{OntologyBase.Size}(interest.Frequency_i)$

    $Score \leftarrow Score * interest.OntologyWeight)$;
    $Return \leftarrow Score$;

Algorithm 3: Check community agent similarity plan

Input: $SourceComProfile$ with the profile community agent source $Agent_O$.
Output: List of recommended communities to $Agent_O$.
begin
    foreach $c \in SourceComProfile$ **do**
        $Interest_i \leftarrow CheckOntologicalBase(c)$;
        foreach $Agent_j \in Communities$ **do**
            $sendComAgent_j(Interest_i)$;

        foreach $Agent_j \in SourceComProfile$ **do**
            $Interest_j \leftarrow receiveComAgent()$;
            $Score_j \leftarrow CalculateCommunInterestScore()$;
            $AG(Interest_i) \leftarrow AG(Interest_i) \cup AG(Interest_j) + Score_j$;

    $Return \leftarrow AG(Interest)$;

Algorithm 4: Computing the new potentials users to a community

---

Input: $SourceComProfile$ with the profile community agent source $Agent_O$.
Output: List of recommended users for the community $Agent_O$.
begin
    foreach $c \in SourceComProfile$ **do**
        $Interest_i \leftarrow CheckOntologicalBase(c)$;
        $sendAllUserAgents(Interest_i)$;
        $Interest_j \leftarrow receiveAllUserAgents()$;
        $Score_j \leftarrow CalculateInterestScoreNewUsers()$;
        $AG(Interest_i) \leftarrow AG(Interest_i) \cup AG(Interest_j) + Score_j$;

Knowledge-Intensive Word Disambiguation
via Common-Sense and Wikipedia

Vládia Pinheiro, Vasco Furtado, Lívio Melo Freire, and Caio Ferreira

Programa de Pós-Graduação em Informática Aplicada – Universidade de Fortaleza (UNIFOR)
Av. Washington Soares, 1321, Fortaleza, Ceará, Brasil
{vladiacelia,vasco}@unifor.br,
{livio.amf,caioferreirax}@gmail.com

Abstract. A promising approach to cope with the challenges that Word Sense Disambiguation brings is to use knowledge-intensive methods. Typically they rely on Wikipedia for supporting automatic concept identification. The exclusive use of Wikipedia as a knowledge base for word disambiguation and therefore the general identification of topics, however, have low accuracy vis-à-vis texts with diverse topics, as can be the case with blogs. This motivated us to propose a method for word disambiguation that, in addition to the use of Wikipedia, uses a common sense database. Use of this base enriches the definition of the concepts previously identified with the help of Wikipedia, and permits the definition of a similarity measure between concepts, which is characterized by verifying the similarity of two concepts from the viewpoint of conceptual proximity in the Wikipedia hierarchy, in addition to the proximity between such concepts in terms of the inferences that they can make. We show that by doing this, we improved the accuracy of automatic disambiguation of words compared with methods that do not use a common sense base.

Keywords: Word Sense Disambiguation, Common Sense Knowledge, Wikipedia.

1   Introduction

Word Sense disambiguation has been challenging Natural Language Processing researchers since the earliest days of Artificial Intelligence (AI). Stevenson and Wilks [1] said that Word Sense Disambiguation (WSD) is an "intermediate task", which is not an end in itself, but rather is necessary at one level or another to accomplish most natural language processing tasks, such as machine translation, information retrieval, hypertext navigation, content and thematic analysis, etc. Recently, the task of automatic topic identification for web content in natural language, such as blog posts, web news etc., requires the comprehension of concepts, sentences, the entire text, and ultimately requires the automatic disambiguation of keyphrases.

A very promising approach to cope with the challenges that WSD brings is to use knowledge-intensive methods. Typically these methods rely on Wikipedia for supporting automatic concept identification and then for supporting disambiguation

L.N. Barros et al. (Eds.): SBIA 2012, LNAI 7589, pp. 182–191, 2012.
© Springer-Verlag Berlin Heidelberg 2012

via the analysis of the proximity of the meaning of the word to other concepts within the text. These methods define similarity measures between concepts, such as the minimum path length in the conceptual hierarchy [2] or the intersection of sets formed by links that are in the article [3].

The exclusive use of Wikipedia as a knowledge base for word disambiguation and therefore the general identification of topics, however, have low accuracy vis-à-vis texts with diverse topics, as can be the case with blogs. This motivated us to propose a method for word disambiguation that, in addition to the use of Wikipedia, uses a common sense database [4], an integral part of the SIM (*Semantic Inferentialism Model*) [5,6]. Use of this base enriches the definition of the concepts previously identified with the help of Wikipedia, and permits the definition of a similarity measure between concepts, which is characterized by verifying the similarity of two concepts from the viewpoint of conceptual proximity in the Wikipedia hierarchy, in addition to the proximity between such concepts in terms of the inferences that they can make. We show that by doing this, we improved the accuracy of automatic disambiguation of words compared with methods that do not use a common sense base.

The article has the following structure. We first formalize the task of knowledge-intensive word disambiguation. This formalization serves as a conceptual framework for the description of related works as well as a description of our proposal that follows. An evaluation of this method is presented as well.

2   Background Knowledge

2.1   Wikipedia as a Repository of Concepts

Wikipedia is used as a concept base for disambiguation. The fundamental element of Wikipedia is the article, which describes named entities, events, and real-world concepts, and contains a set of links to other pages, whether internal or external. The anchor text of links can be seen as a way of referring to the article that is often different from its title. The articles have unique identifiers, which are occasionally complemented with tags that distinguish that page from other pages with the same title, or to reinforce the meaning with which the page is associated. For example, the page of the city of São Paulo has the identifier *São_Paulo_(city)*, in which the "city" tag reinforces the meaning of the article and distinguishes it from ambiguous articles that have *São Paulo* as the title.

Variations in the spelling of a concept or entity – which are commonly present in natural language – are addressed in Wikipedia by the redirection pages. These pages contain redirections to the alternative names that an article might present, such as abbreviations, acronyms, alternate spellings, and even spelling errors. These pages also have identifiers that serve merely as a shortcut to the original article. As it is common to have several articles being referenced by the same term, Wikipedia created disambiguation pages that group together the different meanings that one term may have. These pages are identified by the text combined with the tag "disambiguation". For example, the identifier *Fortaleza_(disambiguation)* refers to a page that groups the articles that have Fortaleza as the title.

The structure of Wikipedia allows the construction of a lexicon that has references to the various meanings of its words. Titles of articles, combined with the anchor text of the articles and the disambiguation/redirection pages, represent changes that reflect the characteristics of polysemy and synonymy that exist in natural language.

2.2   The Common Sense Knowledge Base – InferenceNet

The InferenceNet knowledge bases [4] were constructed containing common sense and inferentialist semantic relations about concepts and sentences, which are expressed in Portuguese and English. InferenceNet is a resource that was constructed according to the Semantic Inferentialist Model (SIM) [5] and expresses the pragmatic character of natural language through pre-conditions and post-conditions of the use of concepts and sentences. According to the inferentialist view [6], the content of a concept must be expressed, becoming explicit, through the use of such concept in inferences, as premises or conclusions of reasoning. Moreover, what determines the use of a concept in inferences or potential inferences in which this concept may participate are: (i) its pre-conditions or premises of use — what gives someone the right to use the concept and what could exclude such a right, serving as premises for utterances and reasoning; and (ii) its post-conditions or conclusions of use — what follows or what are the consequences of using the concept, which let one know what someone is committed to by using a particular concept, serving as conclusions from the utterance per se and as premises for future utterances and reasoning. Formally, InferenceNet's Conceptual Base is represented in a directed graph $G_c(C,Rc)$. Each inferential relationship $rc_j \in Rc$ (set of inferential relations of the concept c) is represented by a tuple (*relationName*, c_i, c_k, *type*), where *relationName* is the name of an InferenceNet semantic relation (*Capableof, PropertyOf, EffectOf* etc.), c_i and c_k are concepts of a natural language, and *type* = "Pre" or "Pos" (pre-condition or post-condition for using the concept c_i). For example, a post-condition of the concept **crime** is (*CapableOf*, "crime", "have victim", "Pos").

SIM defines the inferential relatedness measure $sir(c_i, c_j)$ [5], which expresses how similar two concepts are. According to SIM, the more similar the circumstances and consequences of usage of both concepts are, the more similar the inferences in which the concepts can participate. Accordingly, the more the set of pre-conditions (or post-conditions) of one concept is similar to the set of pre-conditions (or post-conditions) of another concept, the more the two concepts will be inferentially related.

2.3   Formalization of the Process of Knowledge-Intensive Word Disambiguation

The knowledge-intensive disambiguation process relies on knowledge bases such as dictionaries, encyclopedias and thesauri to find – from among the candidate concepts – the one that best expresses the sense or meaning of the word being used in a given context. The input of the disambiguation process is an input text in natural language. The disambiguation process begins with the extraction of the list $P = \{p_1, p_2, ..., p_m\}$ of candidate keyphrases from the input text. To build this list, the document must be tokenized and annotated with part-of-speech and chunk tags, then syntactic filters are used to select phrases belonging to specific grammatical categories, mainly noun phrases [7].

For each phrase $p_i \in P$, we obtain from Wikipedia a list $A_i = \{a_{i1}, a_{i2}, ..., a_{in}\}$ of candidate articles to represent the proper sense of p_i. In some cases, particularly when the keyphrases are very specific, there is no lexicon in Wikipedia that corresponds to the keyphrase. Thus, the keyphrases $p_k \in P$ such that $A_k = \emptyset$ are replaced by their more general set of keyphrases $P_k' = \{p_{k1}', p_{k2}', ..., p_{kr}'\}$, such that $|A_{kj}| > 0$, redefining the set $P = (P - \{p_k\}) \cup P_k'$. For example, the keyphrase "teachers' strike" has no articles for representing its meaning on Wikipedia, so it is replaced by the more general keyphrase "strike".

To form the disambiguation context, the set $C \subseteq P$ is determined, defined as $C = \{p_i \in P \mid |A_i| = 1\}$. The keyphrases $p_i \in C$ have their meanings defined, which are used to carry out the disambiguation of the meanings of ambiguous words. However, as not all keyphrases are semantically relevant to the context in question, it is necessary to prune those keyphrases of little significance and assign weights to the others. Thus, for each $p_i \in C$, a value q_i is calculated, which determines the importance of p_i to the context, according to the following equation:

$$q_i = \sigma_i \times \frac{\sum_{p_j \in C,\ p_j \neq p_i} relatedness(p_i, p_j)}{|C| - 1},$$

where σ_i denotes the value keyphraseness of the keyphrase p_i.

As in [8] and [9], the measure of keyphraseness is used to assign importance to the keyphrases in the text. Because it is a probability *a priori* that a keyphrase is a Wikipedia link, this value is a good indicator to highlight keyphrases that are common topics to the detriment of those that usually appear in the texts just to complete the meaning of the sentences. For example, the keyphrase "Brazil" has keyphraseness close to 50%, since it is often mentioned as a link in Wikipedia articles. However the keyphrase "country" is almost never used as a link and has keyphraseness close to 4%.

The semantic similarity measurement *relatedness*(p_i, p_j) is defined according to the Wikipedia articles that delimit the meanings of the keyphrases p_i and p_j.

Finally, the value m is defined as the maximum size of the disambiguation context. Thus, the context is pruned by removing phrases with lower values of q_i until $|C| = m$. The meanings of resulting keyphrases form the set $S = \{s_1, s_2, ..., s_m\}$, where s_i is the article that defines the meaning of the keyphrase $p_i \in C$.

This context will be useful to determine the best meaning of the keyphrases that make up the set of ambiguous phrases $G = P - C$, comprised of keyphrases that have more than one candidate to represent the meanings thereof. In order to discriminate the article that represents the best meaning of the phrases $p_i \in G$, in relation to the context C, a value w_{ij} is assigned to the items $a_{ij} \in A_i$. Thus, the correct meaning of the phrase p_i is the article a_{ik} with maximum w_{ik} according to the following equation:

$$a_{ij} \mid w_{ij} = \arg\max_{a_{ik} \in A_i} w_{ik}$$

The value w_{ij} defines the semantic similarity between the article a_{ij} and the articles $s_i \in S$, according to the following equation:

$$w_{ij} = \frac{\sum_{s_i \in S} relatedness\left(a_{ij}, s_i\right)}{|S|}$$

The output of the disambiguation process is therefore the definition, for each keyphrase $p_i \in P$, of the article a_{ij} that better expresses the meaning of p_i.

3   Word Disambiguation via Common Sense Knowledge and Wikipedia

In this section, we describe the method developed in order to accomplish the task of automatic disambiguation. We stress that the differential of this study is characterized by the adoption of a semantic relatedness measure that uses common sense knowledge. To develop such a measure, one needs to know which concepts determine the meaning of the Wikipedia articles. For example, for the word *jaguar*, one can find Wikipedia articles referring to the meanings "animal" and "car", among others. For a human reader, the meaning is rapidly clarified in the first paragraph of each article, which – according to Wikipedia's writing guidelines – briefly describes the content in question. These descriptions generally use expressions that establish relations of the "is_a" type. Returning to the example, the car *jaguar* is described by the statement *"Jaguar is a car manufacturer."* The animal *jaguar* is defined by the expression *"The Jaguar is a feline."* From these phrases, one can extract the concepts "car" and "feline", respectively. The disambiguation of the correct meaning for the word *"jaguar"* depends on the context in which it appears. In this context, there are other concepts that have common sense relations with some of the concepts that define the meanings. For example, in a context that has the concept *"carnivore"*, it is expected that the *"animal"* meaning is chosen of the word *jaguar*, because in the common sense base, there should be the relationship *"feline is a carnivore."*

This paper presents a new measure of semantic relatedness that uses knowledge described in Wikipedia articles and a common sense knowledge base. Our argument is that intensive use of common sense knowledge improves the quality of disambiguation of keyphrases.

The basic idea is to identify the related concepts in each Wikipedia article to be compared, and calculate the similarity (or relatedness) between them using the inferential relatedness measure proposed in [5]. The concepts articulated in the description of the article assign the semantic value of the concept per se that is addressed in the article. In Figure 1, for example, we have the opening words of the article that deals with the concept *Jaguar* (animal). In that article, we find the following related concepts: *cat*, *feline*, among others. These concepts, in turn, assign semantic characteristics to the concept *Jaguar*.

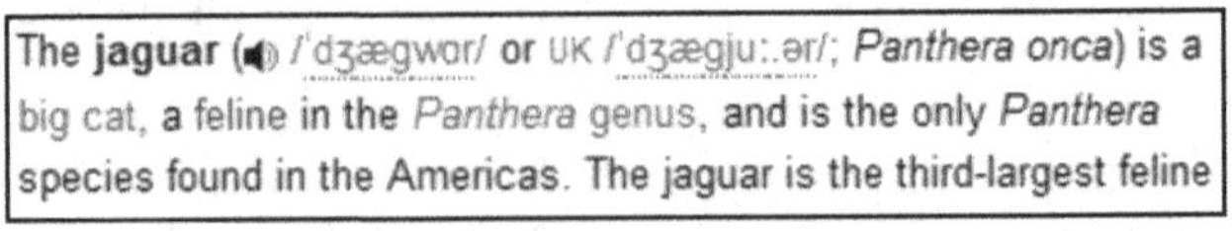

The jaguar (◀) /ˈdʒæɡwɑr/ or UK /ˈdʒæɡjuːˌər/; *Panthera onca*) is a big cat, a feline in the *Panthera* genus, and is the only *Panthera* species found in the Americas. The jaguar is the third-largest feline

Fig. 1. First paragraph of the article *Jaguar* (animal)

The first paragraph of Wikipedia articles is written with the intent of briefly describing the subject matter that follows. In articles about individuals, the first paragraph states the person's dates of birth and death, in addition to place of birth and profession or relevant activity. For places, the first paragraph indicates the type of locality and the city, state and/or country to which it belongs. For organizations, the field of activity and place of origin are defined. For events, the first paragraph refers to the date on which it occurred as well as the location and type of event. For real-world concepts, in general, the articles first present the definition of such concept, other related concepts, and the more generic classes that represent it.

From this semi-structured description, it is possible to extract semantic relations of the "is_a" type. For example, from the introduction of *Dilma Rousseff* (Brazilian president) page, one can extract, among other things, that she is an *economist* and a *politician*. In addition, links that are in the first paragraph of the article highlight important topics of the article and can be considered relevant to the articles in which they appear. The tags, when existing in the identifier of an article, are also used to conceptualize it. For example, to the article whose identifier is *New_Orleans_(film)* is associated the related concept *film*.

In summary, the process of identifying concepts related to a particular article consists of the following steps:

1. Define the *part-of-speech tags* of the texts of the first paragraph of each article. Here, only the phrases consisting of nouns and adjectives are chosen as candidates, according to the results described in [7], reinforcing the idea that, in general, such noun phrases describe the most relevant topics of a text. Additionally, in order to compose the list, the most specific noun phrases are prioritized. In other words, let's say that in the text there is the fragment "the international economy"; the phrase chosen as a candidate is "international economy" instead of the words "economy" and "international".

2. Identify the sentences with the verb "to be" used to predicate a subject that is the same title of the article. In this case, the noun phrases that follow the verb "to be" in the sentence (e.g.: *The jaguar is a big cat*) are defined as concepts related to the article.

3. Define the existing tags on the identifier of the article as related concepts.

4. Define the links that are present in the sentence of the first paragraph and that do not establish relationships of the "is_a" type as related concepts.

5. Based on the set of related concepts, defined in the previous steps, only those belonging to the common sense base are selected. If the concept, as it appears in the description of the article, does not exist in the base, the concept that expresses the sense of the core of the noun phrase is searched.

To illustrate, in Figure 1 we have the first paragraph of the concept *Jaguar (animal)*. According to step 2 above, the semantic relationships "*jaguar is a big cat*" and "*jaguar is a feline*" are extracted. The predicates of the relationships are the concepts "*big cat*" and "*feline*". Because the InferenceNet common sense base does not contain the concept "*big cat*", the concept that represents the core of the noun phrase is searched, which in this case is the concept "*cat*". In the end, the concepts related to the article are "*cat*" and "*feline*".

Through the process described above, each concept defined by the article a_{ij} has a list of related concepts $C_{ij} = \{c_1, c_2, ..., c_t\}$. The measure of semantic relatedness between two articles a_{mn} and a_{kp} is calculated by the following formula, which represents the average of the measure of inferential relatedness $sir(c_i, c_j)$, defined by [5], among all concepts of the sets C_{mn} and C_{kp}, compared among one another:

$$\text{relatedness}(a_{mn}, a_{kp}) = \frac{\sum_{c_i \in C_{mn}} \sum_{c_j \in C_{kp}} sir(c_i, c_j)}{|C_{mn} \times C_{kp}|}$$

4   Empirical Evaluation

In order to validate our proposal, we performed a comparative quantitative evaluation. We used texts from the collaborative portal Wikinews.org (http://pt.wikinews.org/wiki/Página_principal) as our test database. On this site, news articles have links to Wikipedia, which are manually disambiguated. As *Wikinews* is a reputed site maintained by journalists, our basic premise is that this disambiguation is done correctly. We scanned the texts to identify the keyphrases with links, in order to label them as ambiguous or non-ambiguous. A keyphrase is labeled ambiguous if during the search for its meaning in Wikipedia more than one candidate is found. We are then able to measure the accuracy rate of our approach, which consists of scanning the keyphrases labeled ambiguous to find their meaning (the Wikipedia link) and compare them with the existing link in the original *WikiNews* text. If they are the same, then we considered the disambiguation successful.

Following this methodology, we compared our approach with *WikipediaMiner* [10,11], a system that also uses Wikipedia and has good performance in WSD [1]. *WikipediaMiner* also uses rules about the relevance of the keyphrases based on Machine Learning applied to Wikipedia articles. We explored 30 sets of ten randomly selected texts from Wikinews (in Portuguese, because our common sense database is in that language) to count how many correct keyphrase disambiguations each method made. Table 1 shows the results. Normality tests were conducted, which indicated with 98% certainty that the samples follow a normal distribution. In the majority of the rounds (20 out of 30), our approach (*InferenceNet*) outperformed *WikipediaMiner* (*Miner*). Averaging all the test sets and according to Student's test (t =2.1043), it is possible to affirm with 95% confidence that the difference between the accuracy rates of the two approaches is significant, not only due to sampling error.

Table 1. Results of the comparative evaluation

Set	Total Ambiguous	InferenceNet Accuracy (%)	Miner Accuracy (%)	Set	Total Ambiguous	InferenceNet Accuracy (%)	Miner Accuracy (%)
1	38	73.68	73.68	16	36	80.55	80.55
2	34	82.35	76.47	17	34	70.58	64.70
3	36	66.66	63.88	18	25	72.00	72.00
4	23	73.91	60.86	19	22	90.90	77.27
5	35	80.00	71.42	20	25	84.00	80.00
6	29	82.75	79.31	21	32	87.50	78.12
7	29	72.41	72.41	22	22	77.27	72.72
8	49	63.26	61.22	23	34	70.58	70.58
9	28	60.71	50.00	24	24	66.66	58.33
10	24	83.33	79.16	25	33	69.69	72.72
11	27	77.77	77.77	26	36	83.33	80.55
12	53	77.35	73.58	27	29	68.96	68.96
13	46	78.26	67.39	28	41	73.17	73.17
14	35	65.71	65.71	29	25	72.00	76.00
15	29	65.51	55.17	30	36	80.55	72.22
				AVG (%)		75.05	70.86
				SD (%)		7.54	7.88
				t		2.1043	

The results depend on the quality of the common sense database, but the fact that we have conducted the tests randomly gives more credibility to the approach. A qualitative analysis showed that our method, considering the inference ability of the concepts expressed in the common sense dataset, extends the ability to identify the meaning of the algorithm and thus allows better accuracy in disambiguation. In addition, a qualitative analysis of the texts in which *WikipediaMiner* failed shows that this is due to the high diversity of the texts. An emblematic example was when reading a text in which there was mention of a nickname for a football team (which was also the name of a stadium), *WikipediaMiner* failed. In this example we have two ambiguous words but in the same domain. Our approach using the common sense database was able to cope with that.

5 Related Work

Approaches to disambiguation found in the literature are divided into *knowledge-intensive methods* and *machine-learning methods*. In the first strategy, as we have already stated, knowledge bases such as dictionaries, encyclopedias and thesauri are used to find the concepts that best express the sense or meaning of the word being used in a given context. In the second strategy, it is necessary to have a training set that presents words associated with their disambiguated meanings. The success of this approach depends on the selection of attributes able to capture the characteristics that determine the disambiguation criteria. In both methods, there are restrictions for the acquisition of knowledge, whether in the definition of the knowledge repository or in the process of learning from a set of examples.

The strategy defined in [8], called *Wikify!*, combines knowledge-intensive and machine-learning WSD methods in a system of voting. The system extracts words or phrases relevant to the documents and associates each one of them to a Wikipedia article. The knowledge-intensive WSD process is based on the Lesk algorithm [12], which defines the meaning of a word according to the measure based on the overlapping of the dictionary definitions for an ambiguous word and the definitions of the words that make up a disambiguation context (words of the text that have no ambiguity). The part of the method based on machine learning uses features such as part-of-speech tags and a local context defined by the words that are close.

In [13], Wikipedia is used not only as a controlled vocabulary, but also as a measure of similarity (*relatedness*) between items. For each ambiguous word of the text, the value of the semantic relationship between it and the context is calculated according to the links that the articles have in common in the body of their respective texts, as suggested in [3]. Moreover, the popularity *a priori* of articles in Wikipedia (*commonness*) – defined by the ratio between the number of times that a word appears as anchor text of the links to this article and the number of times it appears as anchor text for any article – is also used for the WSD task. A machine-learning algorithm uses the characteristics of *relatedness* and *commonness* to learn how to disambiguate.

Milne and Witten [10] start from the principle that some words comprising the context are more representative than others. They propose assigning weights to the articles that comprise the context of the text based on the semantic relationship of such article to the other articles in the context. The weight given to the semantic relationship of the words in the context is combined with *keyphraseness* measure, defined as the ratio between the number of articles in which a word appears as a link and the number of times such word appears overall, whether as a link or not. In order to disambiguate, a classifier is trained based on the semantic relatedness between the candidates and the context, and the *a priori* popularity of the article.

The approach described in [9] applies a strategy of pruning that removes the least important words from the set used for disambiguation. Therefore, a subset of words is selected, with a predefined size that has the highest *keyphraseness*. The procedure combines the contextual semantic measurements of the ambiguous words with *commonness* measurement.

6   Conclusion

In this paper we presented our approach to working with knowledge-intensive word disambiguation. The method differentiates from the literature due to the use of a common sense dataset and Wikipedia.

We evaluated our approach of word sense disambiguation quantitatively and comparatively with a similar system that uses Wikipedia and machine learning. The results are promising and indicate the advantages of our approach.

The algorithm described in this article is an important component of a platform we have designed to generate widgets that have the ability to identify topics in blog posts and to retrieve contextual information from LOD [14].

Our approach has limitations that will guide our future work. The main point is to improve the similarity measure to analyze the sentence the keyphrase is inserted into. Doing so, we count on capturing the semantic of the words with more precision than we are currently doing.

References

1. Stevenson, M., Wilks, Y.: Word Sense Disambiguation. In: Mitkov, R. (ed.) Oxford Handbook of Computational Linguistics, pp. 249–265. Oxford University Press (2003)
2. Strube, M., Ponzetto, S.: WikiRelate! Computing semantic relatedness using Wikipedia. In: Proceedings of the National Conference on Artificial Intelligence, pp. 14–19. AAAI Press, London (2006)
3. Witten, I., Milne, D.: An effective, low-cost measure of semantic relatedness obtained from Wikipedia links. In: Proceeding of AAAI Workshop on Wikipedia and Artificial Intelligence: An Evolving Synergy, pp. 25–30. AAAI Press, Chicago (2008)
4. Pinheiro, V., Pequeno, T., Furtado, V., Franco, W.: InferenceNet.Br: Expression of Inferentialist Semantic Content of the Portuguese Language. In: Pardo, T.A.S., Branco, A., Klautau, A., Vieira, R., de Lima, V.L.S. (eds.) PROPOR 2010. LNCS (LNAI), vol. 6001, pp. 90–99. Springer, Heidelberg (2010a)
5. Pinheiro, V., Pequeno, T., Furtado, V., Nogueira, D.: Natural Language Processing Based on Semantic Inferentialism for Extracting Crime Information from Text. In: Proceeding of the IEEE Intelligence and Security Informatics (ISI), pp. 19–24 (2010b)
6. Brandom, R.B.: Articulating Reasons: An Introduction to Inferentialism. Harvard University Press, Cambridge (2000)
7. Mihalcea, R., Tarau, P.: TextRank: Bringing order into texts, pp. 404–411. Association for Computational Linguistics, Barcelona (2004)
8. Mihalcea, R., Csomai, A.: Wikify!: Linking documents to encyclopedic knowledge. CIKM 7, 233–242 (2007)
9. Li, C., Sun, A., Datta, A.: A Generalized Method for Word Sense Disambiguation Based on Wikipedia. In: Clough, P., Foley, C., Gurrin, C., Jones, G.J.F., Kraaij, W., Lee, H., Mudoch, V. (eds.) ECIR 2011. LNCS, vol. 6611, pp. 653–664. Springer, Heidelberg (2011)
10. Milne, D., Witten, I.: Learning to link with Wikipedia, pp. 509–518. ACM (2008)
11. Milne, D., Witten, I.: An open-source toolkit for mining Wikipedia. In: Proc. of New Zealand Computer Science Research Student Conference, vol. 9 (2009)
12. Lesk, M.: Automatic sense disambiguation using machine readable dictionaries: how to tell a pine cone from an ice cream cone, pp. 24–26. ACM (1986)
13. Medelyan, O., Witten, I., Milne, D.: Topic indexing with Wikipedia. In: AAAI WikiAI workshop (2008)
14. Santos, H., Furatdo, V., Pinheiro, V., Ferreira, C., Vasconcelos, J.E., Shiki, G.: Widgets baseados em conhecimento advindo de dados referenciados e abertos na Web. In: Proceeding of the XVII WebMedia (2011)

Context-Sensitive ASR for Controlling
the Navigation of Mobile Robots

Gabriel Ferreira Araújo and Hendrik Teixeira Macedo

Computer Science Department, Federal University of Sergipe,
São Cristovão SE 49100000, Brazil
gabrielfa@dcomp.ufs.br, hendrik@ufs.br

Abstract. Automatic Speech Recognition (ASR) is a complex task,
which depends on language, vocabulary and context. In the navigation
control of mobile robots, the set of possible interpretations for a command utterance may be reduced in favor of the recognition rate increase,
if we consider that the robot's work environment is quite defined and
with constant elements. In this paper we propose a contextual model
in addition to the acoustic and language models used by mainstream
ASRs. We provide a whole mobile robot navigation system which use
contextual information to improve the recognition rate of speech-based
commands. Recognition accuracy has been evaluated by Word Information Lost (WIL) metric. Results show that the insertion of a contextual
model provides a improvement around 3% on WIL.

1 Introduction

Human-Robot Interaction (HRI) is a important multidisciplinary research field,
which aims to understand, design, and evaluate every communication form among
humans and robots [5]. HRI applications where humans and robots are placed
within the same room, are usually classified as proximate interaction, which
seems to favor the use of gesture-based or speech-based human-robot interaction kind [2].

Even though multimodal interfaces have been recently proposed in order to
provide more intuitive human-robot proximate interaction [6], we are interested
in speech-based interface issues, due the advantage very well posed by [12]: "according to the diffraction property of audio signal, the sound can bypass obsta-
cles" and so, a speech-based system may suit better for a mobile robot navigation
control, as has been usually applied [13], [1].

In order to be effective, though, speech-based interfaces need high automatic
speech recognition(ASR) rates with low response time to human commands.
Unfortunately, this is still not the case, partially due to language, vocabulary
and context issues. The set of possible pragmatic interpretations for a given
speech command can be reduced through the use of *contextual information* and
so, ASR rates could be raised.

In this paper we analyze the influence of contextual elements in ASR rates for
navigation control of mobile robots. Some efforts have been carried out on the

L.N. Barros et al. (Eds.): SBIA 2012, LNAI 7589, pp. 192–201, 2012.

use of contextual information in speech-based systems [8], [7], [10], however, not providing a proper evaluation of the speech recognition enhancement obtained. Thus, we propose a *contextual model* to ASR systems and analyze the influence of this model in speech-based control system by means of ASR widely adopted evaluation metrics.

A brief introduction to ASR is presented in section 2. The context model proposed is depicted in section 3. Section 4 shows how the context model provided is applied to speech-based control of mobile robot navigation. The influence analysis of the context model in ASR is presented in section 5. We conclude the work in section 6 along with some suggested extensions.

2   Automatic Speech Recognition (ASR)

The job of a speech recognizer is thus to map an acoustic signal to a valid sequence of words, which is valid according to syntactic rules of underlying language.

The first step of the recognition task is to appropriately represent the speech waveform into an acoustic feature vector (or evidence) $X = (x_1, x_2, \cdots, x_t)$. Extraction of the Mel-frequency cepstral coefficients (MFCCs) have been largely used for such representation. MFCC warps the linear spectrum of the speech into a nonlinear scale called Mel which attempts to model the human ear sensitivity. An ASR system can be then mathematically described as a mapping of such acoustic evidence X to a sequence of words $W = (w_1, w_2, \cdots, w_n)$, as defined in equation 1, where the term $P(W|X)$ has been rewritten using Bayes' Rule. Thus, the goal of the recognizer is to select the sequence of words that maximizes the underlying product.

$$\hat{W} = \arg\max_{W \in \omega} P(W|X) = \alpha P(X|W)P(W) \tag{1}$$

where α is the constant factor $1/P(X)$. The estimation of the likelihood $P(X|W)$ term is given by an acoustic model, usually modeled into Hidden Markov Models (HMMs), that translate the input into a sequence of phonemes. The term $P(W)$ is estimated by a language model. One of the most popular is the *Ngram*, which is based on a Markovian assumption that the probability of a given word w_i is approximated by the probability of its observing context of $n - 1$ words.

Finally, once the range of candidate sequence of words $W = (w_1, w_2, \cdots, w_n)$ that maximizes the product is huge, an algorithm is used to perform an efficient search in the state space. This is also known as the ASR "decoding" task.

3   Contextual Model

The contextual model we propose aims to represent the set of constant elements into the robot's work environment, providing semantic information to the navigation system. This model needs to be populated at runtime; inserting, updating and removal operations occurs concurrently with navigation. This ensures data integrity even when the work environment is affected at runtime. The external agent does not interact directly to the contextual model and, as a consequence, the elements that may populate the model must be known in advance.

Let W be the set of constant elements in the robot's work environment. Let V be the set of known elements by the system, such that $V \subseteq W$. Thus, we define the contextual model as

$$\gamma = \{v \in V | v \text{ is into } \omega\} \tag{2}$$

where ω represents the work environment of the robot. The contextual model is the semantic knowledge base used to adapt ASR. The model update and definition of updating criteria shall be treated in initial stages of adaptation process. Elements of contextual model may have n properties $(p_1, p_2, p_3, \ldots p_n)$, e.g., location, visibility, etc. These properties are used as influence factors for the adjustment method or as criteria for the model update. The property *location*, for instance, may be used to set the adaptation level (the closest elements provide higher adaptation) or as a criterion to remove an element from the model (removing the farthest elements).

Adaptation occurs in the language model, which must be adjusted so the ASR is more prone to recognize words semantically related to the elements of the contextual model. Adjustment method is similar to smoothing techniques commonly applied to language models, where probabilities are adjusted by setting new counts c^* for the n-grams. Likewise, in the adjustment method we recalculate the n-gram counts and the probabilities are estimated once more. Let G, a n-gram $w_1 w_2 \ldots w_n$ defined as a set of words $G = \{w_1, w_2, \ldots w_n\}$, and let $\gamma = \{v_1, v_2, v_3, \ldots v_m\}$ be the contextual model populated with m elements. Hence, the set of n-grams from which counts will be increased is defined as follows

$$I = \bigcup_{i=1}^{m} \{G \in Z | G \cap W_{v_i} \neq 0\} \tag{3}$$

where Z is the set of n-grams from textual *corpus* and $W_{v_i} = \{w_1, w_2, w_3, \ldots w_p\}$ is the set of words semantically related to the contextual element v_i. As a result, for each n-gram with at least one word semantically related to some element in the contextual model, its count will be increased. The set of n-grams from which the counts will be decreased may be all n-grams that do not belongs to I, so $D = Z - I$, or just a subset selected by other criteria.

The n-grams counts increase may change with adjustment method criteria. Indeed, they may have different weights to the adjustment. For that reason, the set is separated in subsets (classes of adjustment) of n-grams with the same adjustment weight k and, finally, we define the new count r for the n-grams that will be decreased as follows

$$r = c - \left(c \sum_{i=1}^{l} k_i \right) \tag{4}$$

where c is the n-gram count, k_i is the adjustment weight and the sum of all weights must belongs to interval $[0, 1]$. The new count r for the n-grams that will be increased is defined as

$$r^j = c^j + \left(\frac{c^j}{\sum_{i=1}^{q} c_i^j} \times k_j \times T_d \right) \tag{5}$$

where c^j is the count for the n-grams of class j, q is the number of n-grams in the class j, k_j is the adjustment weight of class j, and T_d is the total count of n-gram that had the count decreased. As a consequence, the counts taken from the decreasing step are redistributed in each one of the l classes, weighted by k_i.

The table 1 shows an adjustment of a 3-gram language model using a contextual model with the elements *"janela"* (the Portuguese version of the word "window") and *"telefone"* (the Portuguese version of the word "phone"). It is possible to see that the n-grams containing the words *"janela"* and *"telefone"* become more likely to occur. In this work, for each element, the set of related words is statically defined and only contains the name of elements. It also could be retrieved in a systematic process by means of a thesaurus, e.g., the WortNet. The adjustment method uses only one adjustment class with the weight $k = 0, 7$, which has been defined empirically.

Table 1. Language model adjustment with a contextual model containing the *"janela"* and *"telefone"* elements. The probabilities are expressed in $\log_{10}$ scale.

1-gram		2-gram		3-gram	
Original	Adjusted	Original	Adjusted	Original	Adjusted
janela $-2,8130$	$-2,1734$	a janela $-1,3954$	$-0,5597$	até a janela $-1,3828$	$-0,5467$
telefone $-2,1051$	$-1,4654$	o telefone $-0,6117$	$-0,1947$	até o telefone $-0,6413$	$-0,2146$
porta $-2,3545$	$-2,8772$	a porta $-0,9326$	$-1,2567$	até a porta $-1,0110$	$-1,3354$
cadeira $-2,6860$	$-3,2089$	a cadeira $-1,2583$	$-1,5802$	até a cadeira $-1,2380$	$-1,5658$

4   Navigation Control of Mobile Robots

We have created a speech-based HRI to the control of mobile robot navigation. Contextual elements are considered and properly identified by the robot during navigation task. In this section we present an overview of the navigation control system's architecture as well as the speech-based HRI main features. We also detail how contextual information is embedded within the system.

4.1   System's Architecture

The application devised has a proximate and real-time HRI: human and robot are placed in the same environment and robot response time to speech commands is immediate. System's architecture is composed of two main nodes: (1) the base station and (2) the robot. There is a video camera attached to the robot and connected to the base station; its role is to detect environment elements that will populate the contextual model. The system's architecture is subdivided into four main features modules: (1) Recognizer, (2) Planner, (3) Controller and (4) Adapter. An overview of the architecture is shown in the Figure 1.

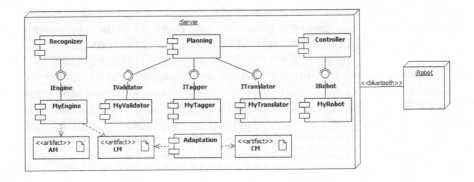

Fig. 1. System's architecture. Main modules are connected to submodules by means of interfaces, making easier the task of editing components.

The Recognizer module encompass the ASR system, and its processing outcome is a set of the most likely sentences $W = (w_1, w_2, \cdots, w_n)$ that maximizes the product of equation 1. Decision-making is done by the Planner module. The requested command is mapped to respective set of robot actions. It consists of three submodules: (a) Validator, (b) Tagger and (c) Translator; The Validator performs the lexicon and syntax validation based on specification of the commands. The Tagger removes words without useful information (such as articles, prepositions, and so on) and replace synonymous words. The Translator maps the recognized sentence to the respective set of robot actions. The role of the Controller module is to provide an interface layer between the base station and the robot, enabling the use of any mobile robotic platform. For each command, the respective actions are delivered consecutively to the robot by means of *bluetooth* connection. In this case study, we have used the robot LEGO Mindstorms NXT 2.0[1]. The Adapter module performs its task concurrently with the main *workflow* described so far, being responsible to provide the adjustment in the ASR system. This module adjusts the speech recognition within the current context, improving the recognition accuracy.

4.2   Speech Command Grammar

Speech recognition has been performed by Julius engine. It is a large vocabulary continuous speech recognition (LVCSR) decoder. In LVCSR systems, the acous-

---

[1] http://mindstorms.lego.com/

tic modeling is focused on the language, regardless the application domain. We have used an open-source HMM acoustic model for Brazilian Portuguese, the LaPSAM [9], which has been built in the HTK software as a cross-word tri-phone model, using 14 Gaussians per mixture and sampling of $22,050Hz$. Language modeling is a particular task in ASR systems because it is defined by the application domain. The Julius engine works with the ARPA format for Ngram languages models and so, we have decided to build our model with the CMU SLM Toolkit, which generates the model from a text corpus. Once the sentence structure of speech commands is fairly simple, we have specified a grammar to generate an automatic text corpus. The main structure of the commands is defined as follows

$$command = \text{``} <s> \text{''}, action, [precision], [reference], \text{``} </s> \text{''} \quad (6)$$

A valid command is composed by a movement action (e.g., walk, turn, go ahead, go back, and so on) which optionally may include precision and reference information. Precision is used to specify some movement parameters, such as distances and angles, whereas the reference specifies directions and location references in the environment, such as left, right, door, window, chair, etc.

Grammar's non-terminal symbols are generated randomly and terminal symbols are generated according to the relative frequency of such symbols (words) into the CETENFolha text *corpus* [11]. CETENFolha is a Brazilian Portuguese text corpus which its sentences have been extracted from texts of the *Folha de São Paulo* newspaper in 1994 (365 issues). The generated *corpus*, called "Simples", has 14.500 transcriptions of different commands, which has been used to build a 3-gram language model by CMU SLM Toolkit, as mentioned earlier.

4.3 Use of Context Information

The contextual model creates a representation of the work environment as described in the section 3. The video camera attached to the robot allows proper identification of environment objects, and thus, the model is populated. Since visual object recognition is not the focus of this work, we simulate this task with fiducial markers (FM), that have been used by imaging systems as reference points and are widely used in augmented reality applications and positioning systems .

We have associated the FM to a set of possible elements in a domestic environment, such as chair, door, wall, table, window, and so on. Each FM detected is immediately inserted into the model and set as visible. Later, the elements whose their respective FM is no longer being detected are set as invisible, but remain in the model. Using the information provided for the contextual model, the adjustment method described earlier is carried out to adjust the language model. In this application we have used a simple adjustment class and weight $k = 0, 7$.

5   Experiments and Results

Recognition rates readings have been analyzed before and after the addition of contextual model. Experiments have been performed using some language models and it also includes validation tests of sentences in the application.

In order to accomplish the recognition tests, we have used four language models n-gram: the Simples model, which has been built from the application domain *corpus* and described in the section 4.2, and the Mixo5k, Mixo10k and Mixo15k models, that have been built by combining the Simples model with 5, 10 and 15 thousands of CETENFolha text *corpus* sentences, respectively. These are 3-gram language models and have been built using the Good-Turing smoothing method [3].

The most common metrics to evaluate n-gram language models are the cross-entropy and perplexity. The entropy measures the uncertainty associated with a random variable whereas the cross-entropy is used to evaluate n-gram language models from a test data set. For a test set T, the cross-entropy H_p is defined as

$$H_p(T) = -\frac{1}{P_T} \log_2 p(T) \tag{7}$$

where P_T is the number of words in T, and $p(T)$ is the probability assigned for the n-gram model to the test set T.

The test set perplexity is also a metric for n-gram models and it is defined by the cross-entropy as follows

$$PP(T) = 2^{H_p(T)} \tag{8}$$

The lower the value of cross-entropy and perplexity, the higher is the model predictive ability. For each one of the four models, we have calculated the cross-entropy and perplexity by a test set consisting of 13.202 words and 2.160 sentences. These sentences have been automatically generated by the grammar. The cross-entropy, perplexity and features of each model are shown in the Table 2.

Table 2. N-gram Models

Models	CETENFolha	Sentences	Vocabulary	Types	Cross-entropy	Perplexity
Simples	—	14.511	closed	79	5, 63	49, 58
Mixo5k	5.000	19.511	open	13.712	2, 01	4, 02
Mixo10k	10.000	24.511	open	20.000	2, 06	4, 17
Mixo15k	15.000	29.511	open	20.000	2, 10	4, 30

For the speech recognition evaluation, we have used a speech corpus composed of 200 utterances: this being 100 positives (valid to the ASR) and 100 negatives (random sentences extracted from CETENFolha text *corpus*).

The most common metric to evaluate ASR systems is the Word Error Rate (WER), which measures the systems' performance in word level by computing the percentual number of substitutions, deletions and insertions required to transform the recognized sentence in the correct one. One of WER's limitations is that its value range is not constant in the interval $[0,1]$ and may produce abnormal results [4]. A better metric is maybe the Word Information Lost (WIL), which has a constant range of values. WIL is defined as follows

$$WIL = 1 - \frac{H^2}{N_1 N_2} \tag{9}$$

where H is the number of correct words (i.e., common to both strings), N_1 and N_2 is the total number of words in each one of strings.

Once the recognition outcomes are indeed affected by the adaptation provided by the contextual model, it is necessary to set higher weights for the language model. The language model weights are usually defined empirically. In the experiments, we have set two different weight values: 20 and 25. The Table 3 shows the WIL rates for all language models, before and after the addition of the contextual model and adjustment method.

Table 3. WIL rates for each language model and weight values of 20 and 25, before and after the addition of contextual model and the adjustment method

	Weight 20		Weight 25	
	Before	After	Before	After
Simples	$24,74\%$	$22,68\%$	$22,53\%$	$19,23\%$
Mixo5k	$37,32\%$	$35,74\%$	$34,12\%$	$33,27\%$
Mixo10k	$39,15\%$	$37,70\%$	$35,69\%$	$36,52\%$
Mixo15k	$38,69\%$	$38,05\%$	$39,11\%$	$37,93\%$

Results show better ASR rates for a weight value of 20 and that the Simples model is the most accurate, since WIL have provided the lowest values. The reduction rate provided by adjustment model has been the highest among the models: above $3,3\%$. This is certainly due to the small and restricted vocabulary, which is specific to the application domain.

We have used the whole speech *corpus* (about 200 utterances) to measure the acceptance of sentences that are recognized by the robot navigation system (i.e., whether sentences are valid or invalid to the submodule Validator). The goal of this test is to evaluate the system's response to invalid commands and noise that may induce to system failure. Binary classifiers can be evaluated by means of ROC space, which crosses true-positive rate ($TPR = \frac{TP}{POS}$) with false-positive rate ($FPR = \frac{FP}{NEG}$). TP is the number of instances correctly classified as positive (accepted), FP is the number of instances wrongly classified as positive, and POS and NEG are the total number of instances correctly and wrongly classified, respectively. The ROC space is a two-dimensional space where the $x-axis$ represents the FPR, the $y-axis$ represents the TPR and each classifier is represented by a point in the space. An ideal classifier would be represented at the point $(0,100)$, where the FPR is zero and the TPR is maximum. As a consequence, the best classifiers are those closest to this point. The Figure 2 shows the representation of models in the ROC space for two weight values, 20 and 25, after the contextual model be added.

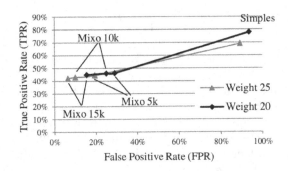

Fig. 2. Language models represented in the ROC space

The validation results show that the Simples model is very distinct from the others, and it is due to the vocabulary type, as previously described in the recognition tests. The closed vocabulary helps to increase the recognition rates and TPR, but in the other hand, these constraints force any sentence match within this vocabulary, which leads to the increase of the FRP.

6   Conclusions

In this paper we have presented a novel approach on the use of context to improve robustness of speech recognition systems. An overview of ASR systems was shown as a brief introduction to the architecture of these systems and their limitations. Under the domain of navigation control of mobile robots, such limitations may increase, mainly due the real-time responsiveness requirement and noisy environment.

We have proposed a context model as an additional component to ASR systems in order to enable recognition adaptation. An adjustment method has been described to accomplish this adaptation within the language model. Furthermore, we have built a speech-based HRI to the task of mobile robot navigation, which includes the context model. We could so evaluate the influence of context model in ASR rates by means of WER and WIL metrics.

Although the vocabulary of language models used in the experiments is restricted, the results demonstrate that the contextual information reduces the set of possible pragmatic interpretations and, thus, improve ASR rates. After introducing the context model populated with some few elements, WIL value has decreased in around 3, 3% in a scenario with a closed and restricted vocabulary.

Future works include carrying out a wider set of experiments with languages models of large vocabulary and use of probabilistic grammars, alternatively to the N-gram models. Also, in order to improve semantic knowledge, we will explore the use of semantic network on the selection of n-grams for adjustment.

References

1. Atrash, A., Kaplow, R., Villemure, J., West, R., Yamani, H., Pineau, J.: Development and validation of a robust speech interface for improved human-robot interaction. Int. J. of Social Robotics 1, 345–356 (2009)
2. Chang, M., Chou, J.: A friendly and intelligent human-robot interface system based on human face and hand gesture. In: IEEE/ASME Int. Conf. on Advanced Intelligent Mechatronics, pp.1856–1861 (2009)
3. Good, I.J.: The Population Frequencies of Species and the Estimation of Population Parameters. Biometrika 40, 237–264 (1953)
4. Green, P., Maier, V., Morris, A.C.: From WER and RIL to MER and WIL: Improved Evaluation Measures for Connected speech Recognition. In: Proc. of 8th Int. Conf. on Spoken Language Processing INTERSPEECH 2004, ICSLP, Jeju, KR, pp. 2765–2768 (2004)
5. Heyer, C.: Human-robot interaction and future industrial robotics applications. In: 2010 IEEE Int. Conf. on Int. Robots and Systems (IROS), pp. 4749–4754 (2010)
6. Maas, J., Spexard, T., Fritsch, J., Wrede, B., Sagerer, G.: BIRON, what's the topic? A Multi-Modal Topic Tracker for improved Human-Robot Interaction. In: The 15th IEEE Int. Symp. on Robot and Human Interactive Commu., pp. 26–32 (2006)
7. Moratz, R., Fischer, K., Tenbrink, T.: Cognitive modeling of spatial reference for human-robot interaction. Int. J. Art. Int. Tools 10(4), 589–611 (2001)
8. Müller, R., Röfer, T., Lankenau, A., Musto, A., Stein, K., Eisenkolb, A.: Coarse Qualitative Descriptions in Robot Navigation. In: Habel, C., Brauer, W., Freksa, C., Wender, K.F. (eds.) Spatial Cognition 2000. LNCS (LNAI), vol. 1849, pp. 265–276. Springer, Heidelberg (2000)
9. Neto, N., Patrick, C., Klautau, A., Trancoso, I.: Free tools and resources for Brazilian Portuguese speech recognition. J. of the Brazilian Comp. Society 17(1), 53–68 (2011)
10. Ross, R., O'Donoghue, R., O'Hare, G.: Improving Speech Recognition on a Mobile Robot Platform through the use of Top-Down Visual Queues. In: Proc. of the 18th Int. Joint Conf. on Artificial Intelligence, IJCAI 2003, pp. 1557–1559 (2003)
11. Santos, D., Bick, E.: Providing Internet access to Portuguese corpora: the AC/DC project. In: Proc. of the Second Int. Conf. on Language Resources and Evaluation (LREC 2000), pp. 205–210 (2000)
12. Sun, H., Yang, P., Liu, Z., Zu, L., Xu, Q.: Microphone array based auditory localization for rescue robot. In: 2011 Chinese Control and Decision Conference (CCDC), pp. 606–609 (2011)
13. Wang, B., Li, Z., Ding, N.: Speech control of a teleoperated mobile humanoid robot. In: 2011 IEEE Int. Conf. on Automation and Logistics (ICAL), pp. 339–344 (2011)

An Evaluation of the Model of Stigmergy in a RoboCup Rescue Multiagent System

Gabriel Rigo da Cruz Jacobsen, Carlos A. Barth, and Fernando dos Santos

Universidade do Estado de Santa Catarina (UDESC), Ibirama, SC, Brazil
gabrielrigoj@gmail.com, {carlos.barth,fernando.santos}@udesc.br

Abstract. A lot of scientists study the behavior of insect's colony like ants, wasps and bees. Through these researches, it is possible to establish patterns used by a group of insects and apply these patterns in other domains. In this paper it will be showed the use of stigmergy in a rescue situation using the RoboCup Rescue simulator. We performed a set of experiments using a metaphor based on the behavior of an ant colony, where the communication between agents is done through the environment. We measured the performance of the ant-based algorithm, expecting to figure out the feasibility of using swarm intelligence in a rescue situation. We compared the results of using stigmergy against a multiagent system based on direct messages. The results showed that the use of stigmergy can outperform the use of direct messages.

1 Introduction

A multiagent system can be described as a system who depends on the relationship of its agents. According to [12], a multiagent system contains agents which interact with each others, and each agent has its own sphere of influence, i.e. any agent has the possibility of being influential in the environment.

The study of social insects has been useful as a way to develop metaphors and have been applied to many areas of knowledge. These applications are present in data analysis, task allocation, optimal search and multiagent systems. One interesting study had been developed by [5] based on the nest reconstruction of termites introducing the concept of stigmergy. Stigmergy can be described by an indirect communication among agents through the environment [2].

The RoboCup Rescue Simulator allows the development of multiagent systems containing agents that can accomplish some specific task in a town destructed by an earthquake. The agents have the ability to communicate among themselves via radio channels, which makes easier calling for help. However, lack of communication caused by the crash of this communication system can directly affect the interactions among agents. A possible solution can be to perform the communication through the environment. However, it has to be noticed this type of communication is not available in the RCR Simulator.

So, this paper deals on two specific points: (i) is it feasible the use of stigmergy in the RCR Simulator? and (ii) is the stigmergy model as efficient as

L.N. Barros et al. (Eds.): SBIA 2012, LNAI 7589, pp. 202–211, 2012.

direct-message approach? To address the second point, we developed a multi-agent system which uses stigmergy, and we compare its performance against a multiagent system which uses direct communication. The results showed that the score of the system based on stigmergy outperforms the score of the system with direct communication by 8.89% depending on the scenario.

The remaining of this paper is organized as follows. Section 2 presents a theoretical introduction to the concept of Swarm Intelligence. Section 3 describes the RCR Simulator. Section 4 shows some related works in the area of multiagent systems and swarm intelligence. Section 5 is about the use of the model of stigmergy in the environment of the RCR Simulator. Section 6 presents the experiments accomplished and an analysis of the results. Section 7 shows the conclusions and future works.

2 Swarm Intelligence

Insect's colony sometimes are described as self-organized systems. Rules specifying interactions among the agents are executed on the basis of purely local information [2]. In the specific case of ants searching for food, the necessary information to led an ant to a food source is a pheromone[1] trail which is left by other ants. Self-organized systems often requires communication between the agents. That communication can be direct or indirect. The indirect communication occurs when an insect changes the environment and after a while other insect perceives that modification and makes a decision based on its perception. This approach of indirect communication is an example of stigmergy. The concept of stigmergy was defined by Grassé [5] when he showed his study about termites rebuilding their nests. He explains that the regulation and coordination of the building activity of termites do not depend only on the workers but is achieved on a cooperative way.

Stigmergy is observed in the process of ant foraging, in which ants search for food. Initially there is no pheromone in the environment, which means the ants take random paths to search for food. When an ant finds a source of food, it moves back to the nest, laying a pheromone trail while walking. When the nestmates sense the pheromone trail, they are stimulated to follow it to the food source. After a while, a lot of ants are engaged in the transportation task, following the shortest pheromone path to the nest.

Deneubourg et al. [3] developed a model of this stigmergy phenomena, where the experimental results matches with the behavior of social insects. The model says the following: the more used a path is, the more attractive it is to an ant, as shown in equation 1. This equation presents a binary path choose where A_i and B_i are the quantities of pheromone dropped in the paths A and B by i ants, respectively.

$$prob_A = \frac{(k + A_i)^n}{(k + A_i)^n + (k + B_i)^n} \qquad prob_B = (1 - prob_A) \qquad (1)$$

---

[1] A pheromone is a chemical used by insects to do communication.

The higher is the amount of pheromone on a path, the higher is the probability of an ant follows that path. The parameter n is used to specify the degree of nonlinearity: when n is high, some path with more pheromone than another, even a slight portion, will take a higher probability of being chosen. The parameter k quantifies the degree of attraction of an unmarked path: when k is high, a higher amount of pheromone is necessary to make the choice nonrandom.

The amount of pheromone on a road A at a timestep t is given by the function $\tau_A(t)$, as shown in 2.

$$\tau_A(t+1) = \rho * \tau_A(t) + \Delta\tau_A(t, t+1) \tag{2}$$

The $\Delta\tau_A(t, t+1)$ is the amount of pheromone dropped between the time t and $t+1$ by every ant i that passed on the path A, as shown in equation 3.

$$\Delta\tau_A(t, t+1) = \sum_{i=1}^{m} \Delta\tau_A^i(t, t+1) \tag{3}$$

One important parameter in the given equation is the decreasement rate ρ. In the real world, it was observed that after a while, the concentration of pheromone in some places has been decreased due to evaporation. It's important to use the coefficient ρ, because when a food source is over, the ants don't have the necessity of going to that place anymore.

3   RoboCup Rescue Simulator

The Great Hanshi-Awaji was an earthquake which hit Kobe City on the 17th of January 1995. This earthquake destroyed approximately 67000 buildings and almost 7000 got burned. More than 4500 people died and nearly 15000 got injuries [1]. After that, the RoboCup Rescue Project started, with the objective of promoting research in multiagent systems, robotics, as well as heterogeneity of agents, planning, information infrastructures and many others [6].

The simulator allows four types of agents: fire brigades, police forces, ambulance teams and civilians. Each agent has its specific functions like extinguishing fire (fire brigades), rescueing civilians (ambulance teams), clearing the streets (police forces) and searching for an ambulance centre (civilians). The agents have limited perception of the environment and must make decisions based on that perception. The communication can be done through a radio channel, and it's possible to choose different configurations which restricts the amount of data that can be sent or the number of channels available. The radio channels have some sort of noise that can mess up the data. In the RCR the noise intensity is a property that can be set as a user-defined probability.

The RoboCup Rescue Simulator is composed by a set of simulators with specific responsibilities. For instance, the traffic simulator manages the locomotion actions of the agents, whereas the collapse simulator manages the debris which obstructs the roads. There is a kernel simulator to control the interactions among all the simulators. The kernel also receives the results obtained by a simulation on each timestep and applies the necessary modifications in the world. When an agent needs to perform an action like move or extinguish, it sends a specific command to the kernel. Every simulation lasts for 300 timesteps and a score is calculated on each one. The score is calculated based on the level of destruction and agents health to measures the performance of the multiagent system.

4 Related Work

In the RCR Simulator, the necessity of a multiagent system which is able to deal with tasks and provide a good score is crucial. Other studies showed that the use of swarm intelligence on the RCR Simulator can lead to good results, furthermore, being efficient. Ferreira Jr. et al. [4] used a probabilistic approach for the agents to choose tasks they will accomplish based on a model inspired by the way the social insects perform task allocation. This algorithm, called Swarm-GAP, has been compared to a greedy task allocation algorithm. The Swarm-GAP showed that the use of swarm-based algorithm was beneficial to the system. Santos and Bazzan [10] also uses a task allocation approach based on extreme teams, named eXtreme-Ants, which is a model to deal with tasks that need to be accomplished by several agents at the same time. These model was compared to other algorithms. The tests have been made in the RCR Simulator and the approach outperformed the other implementations.

As part of the strategy for communication, virtual pheromones can be applied. According [8],[9] and [11] pheromones in an multiagent systems can be virtual as a message protocol or, according [7], a physical component of the system as heat detection. The work of [11] focuses on the capability of the robots in perceiving the environment as a way of making decisions in a collaborative swarm of robots. The work of [9] focuses in a swarm of robots acting in a rescuing scenario. Virtual pheromones are used as a strategy of communication and coordination. In the work of [8] the swarm intelligence is used to give the robots of the swarm the capability of spreading in an unfamiliar environment searching for survivors of a disaster. None of these prior works were proposed in the domain of the RCR Simulator.

5 Stigmergy in the RoboCup Rescue

The RoboCup Rescue was extended to add the following features:

(i) Every road A must have a storage of pheromone τ_A;

(ii) An agent k must be able to sense the amount of pheromone τ_A in the roads of its surroundings;

(iii) An agent k must be allowed to drop an amount of pheromone $\Delta\tau_A^k$ in the road which it is passing;

(iv) The existing pheromone on each road A must evaporate at every timestep by a user-defined value ρ.

In the RCR Simulator every road A has a set of properties which give information about itself. It was added a property called *pheromone* to store the amount of accumulated pheromone τ_A on the whole road. The agents can access the entire set of the road properties, by that way the items (i) and (ii) were accomplished.

When the agents need to move, they send a *move* specific command to the kernel containing the path that must be traveled. So, to meet the (iii) feature it was added a parameter containing the amount of pheromone $\Delta\tau_A^k$ that will be dropped on the whole path. At the end of each timestep, all dropped pheromones are summed up according to equation 2 to give the updated value of τ_A.

The traffic simulator controls the whole traffic present in the simulation, so through this simulator it is possible to access all the roads and modify them. At every timestep these simulator accesses all the roads and decrease the amount of pheromone τ_A by the coefficient ρ, like in the equation 2. This implementation accomplishes the (iv) feature.

6   Experiments and Results

The first experiment we have made aims to validate the RCR extension to support stigmergy. We developed a multiagent system composed only by fire brigades. We decided to disable all types of agents except for the fire brigade agents. This decision is due to the similarity between the fire-fighting actions and the ant-foraging, when ants look for food. When the fire brigade finds a build on fire and its tank of water is out, the agent needs to refuel in the refuge, which is equivalent, in the case of the ants, to move back to the nest. Like an ant releasing pheromone for a specific purpose, this fire brigade agent leaves its own pheromone trail. This trail will serve as a guide back to the building is burning, since the decision is based on the equation 1 (path is chosen based on the amount of pheromones).

To prevent the agents of getting lost, it was developed the following strategy: when an agent is walking for some time without sensing pheromone in any road, the agent must go back to some refuge, so the probability of finding paths with pheromone is higher. The experiments were made in scenarios without either blockages on the roads or civilians.

The experiments were performed in the Kobe4 map (based on the Kobe city) with iniatilly 5 random fire spots and using four different configuration of agents: 20 agents plus 1 refuge (20x1), 20 agents plus 5 refuges (20x5), 40 agents plus 1 refuge (40x1) and 40 agents plus 5 refuges (40x5). The amount of agents and refuges was chosen in a arbitrary way to perform the experiments. In each simulation it was used different amounts of pheromone: no pheromone (zero), 1, 5 and 10 units. It was used also different coeficients of evaporation: 0 (no evaporation), 0.25, 0.50, 0.75 and 1 (total evaporation). These configurations had been chosen to compare if the different quantity of refuges can lead to different results with the same quantity of agents and to compare the score using different amounts of pheromone and coefficient of evaporation. The parameter n of the equation 1 was set accordingly to the best value proposed by [3], which is approximately 2. The parameter k of the same equation 1 was set in 0 to give total emphasis to paths with pheromone. All the results of experiments were averaged over 10 runs of the simulator.

Figure 1 shows the results obtained in the Kobe4 map with 20 agents. As we can see, when the scenario has only one refuge, the score is improved when compared to a scenario with five refuges. One possible way to explain this is due to the fact when the scenario has five refuges, the agents are divided into small groups surrounding their refuges. Otherwise, in the case of only one refuge, a big group is centralized around this refuge. This phenomena occurs because the concentration of pheromone around a refuge is higher than in the first case.

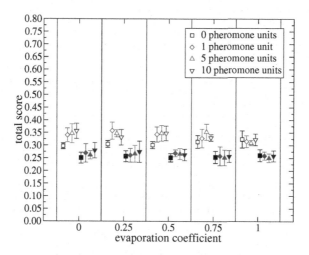

Fig. 1. Results comparing fire brigades using stigmergy in the 20x1 (transparent) and 20x5 (filled) scenarios using different values of pheromone and evaporation

Considering the 20x1 and 20x5 scenarios independently, there is no statistical difference (t test, 95% confidence) in the scores using 1, 5, or 10 pheromone units, meaning that all of these scores were equivalent independently of the coefficient of evaporation (i.e. the scores with 1 pheromone unit are equivalent with all of the evaporation coefficients). However, it is possible to see that when the agents drop some amount of pheromone the results are better than using no pheromone in certain combinations (i.e. with an evaporation coefficient of 0, the use of 10 pheromone units gives a higher score than 0 pheromone units). Since the use of a evaporation coefficient of 1 (total evaporation) is equivalent to using no pheromone, this improvement was not observed in the case of total evaporation. In the 20x1 configuration the use of stigmergy leads to a improvement of 14.35% on average in the score, considering the evaporation coefficients used. In the 20x5 configuration the results were more similar, but an improvement of 11.75% can be seen depending on the configuration.

The Figure 2 presents the results in the same map with 40 agents. The conclusions regarding the map containing 20 agents were not observed here, since there is no statistical difference in the scores when comparing to the map with 1 and 5 refuges. A possible cause is the small quantity of tests that were performed, given that the standard deviation is too high in this map. We also believe it is possible to obtain a better performance by the use of an unevaluated combination of evaporation coefficient and pheromone units.

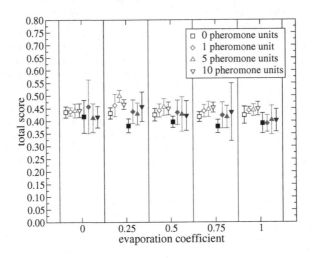

Fig. 2. Results comparing fire brigades using stigmergy in the 40x1 (transparent) and 40x5 (filled) scenarios using different values of pheromone and evaporation

The second experiment compares the performance between two multiagent systems, whose the first one uses the stigmergy model and the second a direct message approach. In a direct-communication-environment, an agent sends a message through a radio channel to all agents while fighting a fire. The message contains a integer (4 bytes) who identifies the building on fire, so the other agents can come to give some help. The simulation creates 3 radio channels with 1024 bytes and with some probabilities used to determine the occurrence of a noise, which can destroy or distort a message sent. All the experiments were performed with the same configuration.

Figure 3 presents the performance of the multiagent systems using stigmergy and direct messages. The results of pheromone agents used in the figure 3 were the best results obtained in the experiments showed in figure 1 and 2.

In the 20x5 configuration, the direct message system has a better performance than the one that uses the stigmergy model. The results of the 20x1 and 40x5 configurations were equivalent. In these cases, the results of the multiagent system which uses stigmergy is as good as the multiagent system which uses direct communication. In the 40x1 configuration the multiagent system which uses stigmergy outperforms the system with direct messages by 8.89%, showing that the use of stigmergy can be better in some cases. We noticed that the higher is the quantity of agents, the higher is the score, due to the quantity of agents to fight

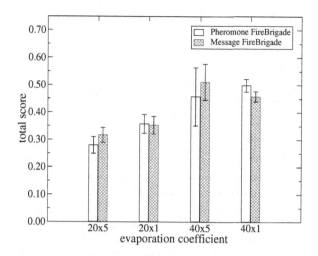

Fig. 3. Results comparing the multiagent systems using stigmergy (Pheromone Fire-Brigade) and direct messages (Message FireBrigade)

fires. When the scenario has only 1 refuge and the agents are using pheromone, the result is better than 5 refuges, which was explained in the hypothesis of small groups and described in the previous experiments. So, the 40x1 scenario would be the best amongst the others, because it combines the two relevant facts which improves the score.

While using messages, the configuration with 5 refuges guarantee a faster refueling to the agents. In the 40x5 configuration, the agents extinguish fires in a big group passing by different refuges. The same pattern is not observed in the 20x5 configuration because the agents are divided into small groups without enough capability of extinguishing a large number of fires. This may occurs because, in the case of 20x5 configuration, the agents need to refuel the tank more times than the agents in the 40x5 configuration. Eventually, in these refueling travels, the agents may find other fire spots and start to divide the group. In the 20x1 configuration, the agents remain in the surroundings of the refuge, extinguishing as many buildings as they can. Comparing the configurations 40x1 and 40x5, the agents seem to have the same behavior. However the distance from the refuge seems to have some influence on the time spent to refuel, reducing the performance of the system in some cases. All the results were based on a t test with 95% confidence.

Regarding our observation, the main issue that impacts negatively on the direct message multiagent system is the use of the communication channel. Table 1 presents the amount of data sent by the agents. Only the results for the direct messages fire brigades were shown, since the fire brigades which use stigmergy does not use the communication channel.

The number of agents and refuges contributes to increase the amount of data to be sent. The data exchange is beneficial because the agents have a large set

Table 1. Use of the communication channel, measured as the average bytes used by every agents in each timestep

scenario configuration	20x5	20x1	40x5	40x1
communication average	27.212 ± 1.451	21.362 ± 1.600	48.283 ± 2.807	33.823 ± 2.894

of possible builds to extinguish, improving the performance of the multiagent system. However, if the communication system was broken, the direct communication would be affected and the agents couldn't send messages. The communication, in this case, could be done through the environment, using pheromones.

Given the use of the communication channel and given the performance of the multiagent system who uses communication (which is better than the use of pheromones in the 20x5 configuration, equivalent in the 20x1 and 40x5, and worse in the 40x1), we can observe that in a disaster scenario with the characteristics of the Kobe4 map and given a broken communication system, the use of stigmergy would be a reasonable solution.

In a real world scenario, the agents could be developed in a way that would be possible to fixate electronic flags along a road. So, other agents could sense that flag and notice that an agent already passed this path. This flag could storage a value of intensity, representing the amount of pheromone. The flag would also control the evaporation based on a configuration.

7   Conclusions and Future Work

There were two main goals in this paper. The first one was to figure out wheter the use of stigmergy in the RoboCup Rescue is feasible or not. The second one aimed a comparison with a model based on direct messaging. The RoboCup Rescue Simulator was extended to support stigmergy. The agents can drop a pheromone trail through the path they are passing and sense the presence of pheromone in the roads. The evaporation is done every timestep to avoid agents going to wrong places. So it is possible now to develop a multiagent system based on stigmergy.

According to the experiments performed, the use of stigmergy can improve the score by 14.35% on average in the case of 20 agents plus 1 refuge. In the case of 20 agents plus 5 refuges the improvement in the score is by 11.77% . Also, there was observed that the performance of using stigmergy can be as good as the model of direct messages in the cases of 40 agents plus 5 refuge and 20 agents plus 5 refuges. The model of stigmergy outperform the model of direct messages in the case of 40 agents plus 1 refuge.

Regarding our research, future works can include more than one sort of pheromone, so that other sorts of agents can be added in the environment. With only one sort of pheromone, the use of various agents would be inconsistent, since each pheromone would be associated with one type of agent. Another approach to be worked is the adaptation of the algorithm of Santos and Bazzan [10] to support the use of stigmergy, so these approaches would focus in two uses of swarm intelligence: task allocation and to communication.

References

1. The great hanshin-awaji earthquake statistics and restoration progress (January 2012), http://www.city.kobe.lg.jp/safety/hanshinawaji/revival/promote/january.2012.pdf
2. Bonabeau, E., Theraulaz, G., Dorigo, M.: Swarm Intelligence: From Natural to Artificial Systems. Oxford University Press, New York (1999)
3. Deneubourg, J.L., Aron, S., Goss, S., Pasteels, J.M.: The self-organizing exploratory pattern of the argentine ant. Journal of Insect Behavior 3, 159–168 (1990)
4. Ferreira Jr., P.R., dos Santos, F., Bazzan, A.L.C., Epstein, D., Waskow, S.J.: Robocup rescue as multiagent task allocation among teams: experiments with task interdependencies. Autonomous Agents and Multi-Agent Systems 20, 421–443 (2010)
5. Grasse, P.-P.: La reconstruction du nid et les coordinations interindividuelles chez bellicositermes natalensis et cubitermes sp. la theorie de la stigmergie: essai dinterpretation du comportement des termites constructeurs. Insectes Sociaux 6(1), 41–80 (1959)
6. Kitano, H., Tadokoro, S., Noda, I., Matsubara, H., Takahashi, T., Shinjou, A., Shimada, S.: Robocup rescue: search and rescue in large-scale disasters as adomain for autonomous agents research. In: Proceedings of the IEEE International Conference on Systems, Man, and Cybernetics (SMC), Tokyo, Japan, vol. 6, pp. 739–743. IEEE (October 1999)
7. Mayet, R., Roberz, J., Schmickl, T., Crailsheim, K.: Antbots: A Feasible Visual Emulation of Pheromone Trails for Swarm Robots. In: Dorigo, M., Birattari, M., Di Caro, G.A., Doursat, R., Engelbrecht, A.P., Floreano, D., Gambardella, L.M., Groß, R., Şahin, E., Sayama, H., Stützle, T. (eds.) ANTS 2010. LNCS, vol. 6234, pp. 84–94. Springer, Heidelberg (2010)
8. Payton, D., Daily, M., Estowski, R., Howard, M., Lee, C.: Pheromone robotics. Autonomous Robots 11, 319–324 (2001)
9. Payton, D., Estkowski, R., Howard, M.: Compound behaviors in pheromone robotics. Robotics and Autonomous Systems 44(3–4), 229–240 (2003)
10. dos Santos, F., Bazzan, A.L.C.: Towards efficient multiagent task allocation in the robocup rescue: a biologically-inspired approach. Autonomous Agents and Multi-Agent Systems 22, 465–486 (2011)
11. Simonin, O., Charpillet, F.: Indirect cooperation between mobile robots through an active environment. In: 5th National Conference on "Control Architecture of Robots", pp. 71–80 (2010)
12. Wooldridge, M.J.: An Introduction to MultiAgent Systems. John Wiley & Sons, Chichester (2002)

An Ecology-Based Heterogeneous Approach for Cooperative Search

Rafael Stubs Parpinelli[1,2,*] and Heitor Silvério Lopes[2]

[1] Applied Cognitive Computing Group
Santa Catarina State University
Joinville, Brazil
[2] Bioinformatics Laboratory
Federal Technological University of Paraná
Curitiba, Brazil
parpinelli@joinville.udesc.br, hslopes@utfpr.edu.br

Abstract. The concept of optimization is present in several natural processes such as the evolution of species, the behavior of social groups and the ecological relationships of different animal populations. This work uses the concepts of habitats, ecological relationships and ecological successions to build a hybrid cooperative search algorithm, named ECO. The Artificial Bee Colony (ABC) and the Particle Swarm Optimization (PSO) algorithms were used in the experiments where benchmark mathematical functions were optimized. Results were compared with ABC and PSO running alone, and with both algorithms in a well known island model with ring topology, all running without the ecology concepts previously mentioned. The ECO algorithm performed better than the other approaches, especially as the dimensionality of the functions increase, possibly thanks to the ecological interactions (intra and inter-habitats) that enabled the co-evolution of populations. Results suggest that the ECO algorithm can be an interesting alternative for numerical optimization.

Keywords: optimization, cooperative search, co-evolution, habitats, ecological relationships, ecology.

1 Introduction

The nature has always been an endless source of inspiration for computational models and paradigms, in particular for the computer scientists of the area known as Natural Computing [1]. The two main families of bio-inspired algorithms are evolutionary computation (EC) and swarm intelligence (SI) that currently offer a wide range of strategies for optimization [2][3].

The concept of optimization is present in several natural processes such as in the evolution of the species, in the behavior of social groups, in the dynamics

* Authors would like to thank the Brazilian National Research Council (CNPq) for the research grant to H.S. Lopes; as well as to UDESC (Santa Catarina State University) and FUMDES program for the doctoral scholarship to R.S. Parpinelli.

of the immune system, in the strategies of search for food and in the ecological relationships of different populations. Most of these cases were the source of inspiration to the development of algorithms for optimization.

It is worth mentioning that most bio-inspired algorithms only focus on and take inspiration from specific aspects of the natural phenomena. However, in nature, biological systems are interlinked to each other, e.g. biological ecosystems [4][5]. Hence, hybrid bio-inspired systems arise and are developed with and defined by cooperative search concepts. Cooperative search strategies involve concepts of parallelism and hybridism where a set of potentially good algorithms for the optimization problem are executed in parallel, sharing information during the run [6]. Some related works have shown good results when using hybrid bio-inspired search strategies cooperatively [7][8][9].

With such diversity of search strategies and the advantages of applying them cooperatively, it is possible to establish an analogy with the dynamics of biological ecosystems. In this analogy with biological ecosystems each species can behave according to a specific search strategy, employed in the evolution of candidate solutions. In [10] the authors first illustrate the potentiality of some ecological concepts (e.g., habitats, ecological relationships and ecological successions) presenting a simplified ecological-inspired algorithm. In this work we explore the heterogeneity of the model using two different algorithms cooperatively: the ABC [11] and the PSO [12]. In addition to the possibility of using different search strategies cooperatively, the methodology opens the possibility for the insertion of ecological concepts in the optimization process, thus bringing more biological plausibility to the system.

2   The Proposed Ecological-Inspired Approach

The ecological-inspired algorithm, named ECO, represents a new perspective to develop cooperative search algorithms. The ECO is composed by populations of individuals (candidate solutions for a problem being solved) and each population evolves according to a search strategy. In this way, the individuals of each population are modified according to the mechanisms of intensification and diversification, and the initial parameters, specific to the search strategy being employed. The ecological-inspired system can be modeled in two ways: homogeneous or heterogeneous. A homogeneous model implies that all populations evolve in accordance to the same optimization strategy, configured with the same parameters. Any change in the strategies or parameters in at least one population characterises a heterogeneous model.

The ecological inspiration stems from the use of some ecological concepts, such as: habitats, ecological relationships and ecological successions [4][5]. Once dispersed in the search space, populations of individuals established in the same region constitute an ecological habitat. For instance, in a multimodal hyper-surface, each peak can become a promising habitat for some populations. A hyper-surface may have several habitats. As well as in nature, populations can move around through all the environment. However, each population may belong only to one habitat in a given moment of time t. Therefore, by definition, the intersection between all habitats at moment t is the empty set.

With the definition of habitats, two categories of ecological relationships can be defined. Intra-habitats relationships that occur between populations inside each habitat, and inter-habitats relationships that occur between habitats [4][5].

In ECO, the intra-habitat relationship is the mating between individuals. Populations belonging to the same habitat can establish a reproductive link between their individuals, meshing the populations and favoring the co-evolution of the involved populations. Populations belonging to different habitats are called reproductively isolated.

The inter-habitats relationship is the great migrations. Individuals belonging to a given habitat migrate to other habitats aiming at identifying promising areas for survival and mating.

In addition to the mechanisms of intensification and diversification specific to each search strategy, when considering the ecological context of the proposed algorithm, the intra-habitats relationships are responsible for intensifying the search and the inter-habitats relationships are responsible for diversifying the search.

Inside the ecological metaphor, the ecological successions represent the transformational process of the system. In this process, populational groups are formed (habitats), relations between populations are established and the system stabilizes by means of the self-organization of its components.

Algorithm 1 shows the pseudo-code of the proposed approach. First of all, at instant $t = 0$, all individuals of all populations Q_i, with $i = 1, \ldots, NQ$, are randomly initialized (line 2). NQ is the total number of populations.

The main loop represents the ecological successions (lines 3 to 12). The first step inside the main loop is the evolutive period for each population $Q_i(t)$ (line 4). In this step each population explores the search area freely in accordance to its intensification and diversification criteria.

Algorithm 1. Pseudo-code for ECO

1: Consider $i = 1, \ldots, NQ$, $j = 1, \ldots, NH$ and $t = 0$;
2: Initialize each population $Q_i(t)$ with n_i random candidate solutions;
3: **while** stop criteria not satisfied **do** {Ecological succession cycles}
4:    Perform evolutive period for each population $Q_i(t)$;
5:    Apply metric C_i to identify the region of reference for each population $Q_i(t)$;
6:    Using the C_i values, define the NH habitats;
7:    For each habitat $H_j(t)$ define the communication topology $CT_j(t)$ between populations $Q_i^j(t)$;
8:    For each topology $CT_j(t)$, perform interactions between populations $Q_i^j(t)$;
9:    Define communication topology $TH(t)$ between $H_j(t)$ habitats;
10:    For $TH(t)$ topology, perform interactions between $H_j(t)$ habitats;
11:    Increase t;
12: **end while**

At the end of the evolutive period of all populations it is necessary to identify the region of reference for each population (line 5), which is used to define the habitats of the system. The metric chosen to define the region of reference is the centroid C_i, with $i = 1, \ldots, NQ$. The centroid represents the point in the space where there is a longest concentration of individuals of population i and is calculated by Equation 1, where Pop_i represents the number of individuals in the population i and x_k represents an individual (solution vector).

$$C_i = \frac{\sum_{k=1}^{Pop_i} x_k}{Pop} \tag{1}$$

Once found the centroids, the distances between populations are calculated and the habitats $H_j(t)$ are defined using a minimum distance threshold $\rho \in [0..1]$ (line 6). The distance between populations is dependent on the definition and representation of each population. In this application we used the Euclidean distance. At the end of this step all populations are associated with a habitat: $Q_i^j(t)$, population i belongs to the habitat j at the moment t.

With the NH habitats defined, it is necessary to determine the intra-habitats communication topologies $CT_j(t)$ for each habitat $H_j(t)$ with $j = 1, \ldots, NH$ (line 7). The topologies are defined by an adjacency matrix between the populations belonging to each habitat. Once defined the intra-habitats topologies, the mating ecologic relationship between adjacent populations occurs (line 8). In this relationship, one individual of each population is chosen using the tournament strategy [13] and a genetic exchange between them is performed in order to generate a new individual. The new generated individual replaces an individual selected at random in its population, excluding the best individual.

Once the interactions between the populations of each habitat have been done, the $TH(t)$ topology for interaction among habitats (line 9) is randomly defined. This inter-habitats topology $TH(t)$ is used for the completion of the great migrations ecologic relationship. In this relationship, for each habitat a random population belonging to him is chosen. The best individual of this population migrates to another habitat and, in the destination habitat, it replaces an individual randomly chosen, excluding the best individual (line 10).

The main loop continues until the ecological succession cycles reach the maximum predefined value.

3   Experiments and Results

Experiments were conducted using four benchmark functions extensively used in the literature for testing optimization methods [14]. Each function was tested with 2, 5, 10, and 200 dimensions and, for all of them, the objective is to minimize.

The first function ($f_1(x)$) is known as generalized F6 Schaffer function [15] that is strongly multimodal. The second function ($f_2(x)$) is the Rastrigin function and it is also multimodal and based on the sphere function with the addition of a cosinoid modulation to produce a large number of local minima [16]. The

Table 1. Benchmark Functions

Function	Definition	Domain	Global Optimun
$f_1(\boldsymbol{x})$	$\sum_{i=1}^{n-1}\left(0.5 + \frac{\sin^2\left(\sqrt{x_{i+1}^2+x_i^2}\right)-0.5}{\left(0.001\left(x_{i+1}^2+x_i^2\right)+1\right)^2}\right)$	$-100 \leq x_i \leq 100$	$f_1(\boldsymbol{0}) = 0$
$f_2(\boldsymbol{x})$	$\sum_{i=1}^{n}(x_i^2 - 10\cos(2\pi x_i) + 10)$	$-5.12 \leq x_i \leq 5.12$	$f_2(\boldsymbol{0}) = 0$
$f_3(\boldsymbol{x})$	$\frac{1}{4000}\left(\sum_{i=1}^{n} x_i^2\right) - \left(\prod_{i=1}^{n}\cos\left(\frac{x_i}{\sqrt{i}}\right)\right) + 1$	$-600 \leq x_i \leq 600$	$f_3(\boldsymbol{0}) = 0$
$f_4(\boldsymbol{x})$	$\sum_{i=1}^{n-1}(100(x_{i+1} - x_i^2)^2 + (x_i - 1)^2)$	$-30 \leq x_i \leq 30$	$f_4(\boldsymbol{1}) = 0$

third function ($f_3(\boldsymbol{x})$) is the Griewank function [17] that, as the other, is strongly multimodal. Particularly the number of local minima for this function increases exponentially with the dimensionality [18], which makes it a very difficult optimization problem. The fourth function ($f_4(\boldsymbol{x})$) is the Rosenbrock function [19], which has the particularity of having a search hyper-surface in the form of a long and narrow parabolic valley, where the global optimum is located. Table 1 summarizes the informations about the functions used.

The parameters of the ECO algorithm are: number of populations (N-POP) that will be co-evolved, number of cycles for ecological successions (ECO-$STEP$), the size of the evolutive period (EVO-$STEP$) that represents number of function evaluations in each ECO-$STEP$, the tournament size (T-$SIZE$) and the proximity threshold ρ. Studies about the adjustment of parameters have not been carried out yet. Hence, all the parameters of the algorithm were defined empirically.

In all experiments the ABC algorithm and the PSO algorithm were used heterogeneously, i.e. half population was managed by the ABC algorithm and the other half was managed by the PSO algorithm to evolve their candidate solutions. For the ABC algorithm, there is only one control parameter, the population size (POP-$SIZE$). For the PSO algorithm, besides POP-$SIZE$, the parameters were set to standard values [20]: inertia weight $W = 0.721$ and confidence factors $C_1 = C_2 = 1.193$.

For the number of dimensions equal to 2, 5 and 10, the parameters used were N-$POP = 100$, POP-$SIZE = 10$, ECO-$STEP = 100$, EVO-$STEP = 100$, T-$SIZE = 5$ e $\rho = 0,5$. With this configuration, the total number of function evaluations is 10,000 for each population. For the number of dimensions equal to 200, the number of populations was redefined to 200, the number of ecological succession cycles was redefined to 500, and the evolutionary period was redefined to 200. With this adjustment of parameters, for 200 dimensions, the total number of function evaluations has become from 10,000 to 100,000 evaluations for each population. All parameters' values were chosen empirically.

3.1   Numerical Optimization of Benchmark Functions

The ecological-inspired algorithm (ECO) was tested using three configurations. The first configuration implements the Algorithm 1 as described in Section 2, with the definitions of habitats, topologies and ecological relations. In the second configuration, the ability to create habitats is disabled and, consequently, topologies and interactions are not defined. This second configuration simulates the evolution completely isolated populations, and they evolve without exchanging information. The third configuration disables the ability to create habitats, but explores co-evolution using multi-populations in a ring topology. In this configuration migration policy and frequency occurs exactly as in the first configuration. For each configuration, the algorithm was run 30 times.

Tables 2, 3, 4 and 5 show the averaged results obtained for the benchmark functions. Each table lists the dimensions and for each dimension the results obtained by each configuration of the algorithms are presented (column 2 to 9). Column 2 shows the results obtained by the ABC algorithm running alone, without co-evolution. Column 3 shows the results obtained by the ABC algorithm running with a ring topology (ABC_{Ring}). Column 4 shows the results obtained by the ABC algorithm using the ecological-inspired approach homogeneously (ECO_{ABC}). Column 5 shows the results obtained by the PSO algorithm running alone, without co-evolution. Column 6 shows the results obtained by the PSO algorithm running with a ring topology (PSO_{Ring}). Column 7 shows the results obtained by the PSO algorithm using the ecological-inspired approach homogeneously (ECO_{PSO}). Column 8 shows the results obtained by the ABC and PSO algorithms alternated within a ring topology (PSO-ABC_{Ring}). In this configuration, half population is controlled by the ABC algorithm and the other half is controlled by the PSO algorithm. Finally, column 9 shows the results obtained by the ABC and PSO algorithms using the ecological-inspired approach heterogeneously ($ECO_{PSO-ABC}$). Again, half population is controlled by the ABC algorithm and the other half by the PSO. For each dimension, the third line (*Global Best*) shows the average and standard deviation of the best result obtained by all populations in all runs.

Table 2. Obtained results for function $f_1(x)$

$f_1(x)$				$D = 2$				
Model	ABC	ABC_{Ring}	ECO_{ABC}	PSO	PSO_{Ring}	ECO_{PSO}	PSO-ABC_{Ring}	$ECO_{PSO-ABC}$
Global Best	$10^{-05} \pm 0.0$	$10^{-15} \pm 0.0$	$10^{-16} \pm 0.0$	0.0000 ± 0.0	0.0000 ± 0.0	0.0000 ± 0.0	0.0000 ± 0.0	0.0000 ± 0.0
					$D = 5$				
Model	ABC	ABC_{Ring}	ECO_{ABC}	PSO	PSO_{Ring}	ECO_{PSO}	PSO-ABC_{Ring}	$ECO_{PSO-ABC}$
Global Best	0.1829 ± 0.0	0.0024 ± 0.0	0.0201 ± 0.0	$10^{-12} \pm 0.0$	$10^{-17} \pm 0.0$	$10^{-11} \pm 0.0$	$10^{-11} \pm 0.0$	$10^{-08} \pm 0.0$
					$D = 10$				
Model	ABC	ABC_{Ring}	ECO_{ABC}	PSO	PSO_{Ring}	ECO_{PSO}	PSO-ABC_{Ring}	$ECO_{PSO-ABC}$
Global Best	4.6569 ± 0.8	1.0974 ± 0.5	1.1344 ± 0.2	0.1484 ± 0.1	0.0107 ± 0.0	0.0604 ± 0.1	0.0101 ± 0.0	0.0053 ± 0.0
					$D = 200$				
Model	ABC	ABC_{Ring}	ECO_{ABC}	PSO	PSO_{Ring}	ECO_{PSO}	PSO-ABC_{Ring}	$ECO_{PSO-ABC}$
Global Best	27.5936 ± 0.7	26.5936 ± 0.1	20.2792 ± 0.4	20.2355 ± 0.0	11.6461 ± 0.3	10.3316 ± 0.3	6.8273 ± 0.3	0.0906 ± 0.0

Analyzing the results for $D = 2$, 5 and 10, *Global Best* line for each dimension, we can observe that the ECO approaches (ECO_{ABC}, ECO_{PSO} and $ECO_{PSO-ABC}$) obtained competitive results when compared to the other corresponding configurations (ABC and ABC_{ring}, PSO and PSO_{ring}, and PSO-ABC_{ring}, respectively), for all functions.

In Figure 1 we can visually verify the results for $D = 200$, where the x-*axis* shows the different approaches and the y-*axis* represents the *Global Best* values of each approach and are shown at the top of each bar. Analyzing these results

Table 3. Obtained results for function $f_2(x)$

$f_1(x)$	$D = 2$							
Model	ABC	ABC_{Ring}	ECO_{ABC}	PSO	PSO_{Ring}	ECO_{PSO}	$PSO\text{-}ABC_{Ring}$	$ECO_{PSO-ABC}$
Global Best	0.0000 ± 0.0	0.0000 ± 0.0	0.0000 ± 0.0	0.0000 ± 0.0	0.0000 ± 0.0	0.0000 ± 0.0	0.0000 ± 0.0	0.0000 ± 0.0
	$D = 5$							
Model	ABC	ABC_{Ring}	ECO_{ABC}	PSO	PSO_{Ring}	ECO_{PSO}	$PSO\text{-}ABC_{Ring}$	$ECO_{PSO-ABC}$
Global Best	0.0000 ± 0.0	0.0000 ± 0.0	0.0000 ± 0.0	0.0000 ± 0.0	0.0000 ± 0.0	0.0000 ± 0.0	0.0000 ± 0.0	0.0000 ± 0.0
	$D = 10$							
Model	ABC	ABC_{Ring}	ECO_{ABC}	PSO	PSO_{Ring}	ECO_{PSO}	$PSO\text{-}ABC_{Ring}$	$ECO_{PSO-ABC}$
Global Best	$10^{-11} \pm 0.0$	0.0000 ± 0.0	0.0000 ± 0.0	1.8904 ± 0.5	1.0944 ± 1.2	0.1014 ± 0.3	0.0000 ± 0.0	$10^{-15} \pm 0.0$
	$D = 200$							
Model	ABC	ABC_{Ring}	ECO_{ABC}	PSO	PSO_{Ring}	ECO_{PSO}	$PSO\text{-}ABC_{Ring}$	$ECO_{PSO-ABC}$
Global Best	62.1453 ± 9.6	18.3877 ± 6.2	$10^{-05} \pm 0.0$	442.4012 ± 8.7	328.7323 ± 12.1	159.7621 ± 0.5	$10^{-06} \pm 0.0$	0.0000 ± 0.0

Table 4. Obtained results for function $f_3(x)$

$f_1(x)$	$D = 2$							
Model	ABC	ABC_{Ring}	ECO_{ABC}	PSO	PSO_{Ring}	ECO_{PSO}	$PSO\text{-}ABC_{Ring}$	$ECO_{PSO-ABC}$
Global Best	$10^{-19} \pm 0.0$	0.0000 ± 0.0	0.0000 ± 0.0	0.0000 ± 0.0	0.0000 ± 0.0	0.0000 ± 0.0	0.0000 ± 0.0	0.0000 ± 0.0
	$D = 5$							
Model	ABC	ABC_{Ring}	ECO_{ABC}	PSO	PSO_{Ring}	ECO_{PSO}	$PSO\text{-}ABC_{Ring}$	$ECO_{PSO-ABC}$
Global Best	$10^{-12} \pm 0.0$	$10^{-18} \pm 0.0$	$10^{-19} \pm 0.0$	0.0000 ± 0.0	0.0000 ± 0.0	0.0000 ± 0.0	0.0000 ± 0.0	0.0000 ± 0.0
	$D = 10$							
Model	ABC	ABC_{Ring}	ECO_{ABC}	PSO	PSO_{Ring}	ECO_{PSO}	$PSO\text{-}ABC_{Ring}$	$ECO_{PSO-ABC}$
Global Best	$10^{-06} \pm 0.0$	$10^{-17} \pm 0.0$	$10^{-13} \pm 0.0$	0.0000 ± 0.0	0.0000 ± 0.0	0.0000 ± 0.0	0.0000 ± 0.0	0.0000 ± 0.0
	$D = 200$							
Model	ABC	ABC_{Ring}	ECO_{ABC}	PSO	PSO_{Ring}	ECO_{PSO}	$PSO\text{-}ABC_{Ring}$	$ECO_{PSO-ABC}$
Global Best	$10^{-7} \pm 0.0$	$10^{-09} \pm 0.0$	$10^{-11} \pm 0.0$	1.6673 ± 0.1	$10^{-16} \pm 0.$	$10^{-18} \pm 0.0$	$10^{-16} \pm 0.0$	0.0000 ± 0.0

Table 5. Obtained results for function $f_4(x)$

$f_1(x)$	$D = 2$							
Model	ABC	ABC_{Ring}	ECO_{ABC}	PSO	PSO_{Ring}	ECO_{PSO}	$PSO\text{-}ABC_{Ring}$	$ECO_{PSO-ABC}$
Global Best	$10^{-08} \pm 0.0$	$10^{-05} \pm 0.0$	$10^{-09} \pm 0.0$	0.0000 ± 0.0	0.0000 ± 0.0	$10^{-17} \pm 0.0$	0.0000 ± 0.0	$10^{-17} \pm 0.0$
	$D = 5$							
Model	ABC	ABC_{Ring}	ECO_{ABC}	PSO	PSO_{Ring}	ECO_{PSO}	$PSO\text{-}ABC_{Ring}$	$ECO_{PSO-ABC}$
Global Best	0.0014 ± 0.0	0.0054 ± 0.0	0.0001 ± 0.0	0.0006 ± 0.0	0.0001 ± 0.0	0.0066 ± 0.0	$0.0001 \pm 0.$	0.0002 ± 0.0
	$D = 10$							
Model	ABC	ABC_{Ring}	ECO_{ABC}	PSO	PSO_{Ring}	ECO_{PSO}	$PSO\text{-}ABC_{Ring}$	$ECO_{PSO-ABC}$
Global Best	0.0098 ± 0.0	0.0184 ± 0.0	0.0086 ± 0.0	0.0317 ± 0.0	0.0695 ± 0.0	0.0350 ± 0.0	0.0757 ± 0.0	0.0034 ± 0.0
	$D = 200$							
Model	ABC	ABC_{Ring}	ECO_{ABC}	PSO	PSO_{Ring}	ECO_{PSO}	$PSO\text{-}ABC_{Ring}$	$ECO_{PSO-ABC}$
Global Best	13036.1 ± 4193.4	139.4325 ± 7.4	137.86 ± 42.0	75089.9 ± 9999.6	247.041 ± 13.1	243.14 ± 2.6	156.0453 ± 16.4	138.25 ± 37.9

we can observe that the homogeneous ecological-inspired approaches, ECO_{ABC} and ECO_{PSO}, obtained better results than the algorithms executed without the concepts of habitats for all functions. Analyzing the results for the heterogeneous ecological-inspired approach, $ECO_{PSO-ABC}$, we can observe that the results were significantly better than the heterogeneous island model with ring topology ($PSO\text{-}ABC_{Ring}$). This gain is mainly due to the ecological interactions (intra and inter-habitats) that favor the co-evolution of populations. Moreover, the $ECO_{PSO-ABC}$ was the best approach for all functions.

Figure 2 shows the evolution of the average number of habitats for each ecological succession step for 30 executions of the ECO_{ABC} algorithm. As the observed behavior is the same for all functions and eco-models, Figure 2 shows the evolution for the function $f_1(x)$ with $D = 10$. It can be observed that, at the beginning of the optimization process, with the populations widely dispersed in the search space, there is a greater number of habitats. To the extent that the optimization process moves through the ecological successions, the populations tend to move through the search space converging to specific regions. As shown in Figure 2,

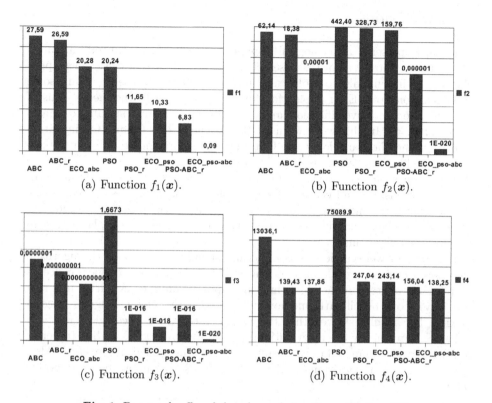

(a) Function $f_1(x)$.

(b) Function $f_2(x)$.

(c) Function $f_3(x)$.

(d) Function $f_4(x)$.

Fig. 1. Bar graph off each benchmarck function with $D = 200$

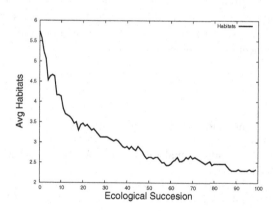

Fig. 2. Average number of habitats at each ecological succession

the number of habitats decreases with the ecological succession cycles, indicating that the populations tend to converge to points close to each other.

4   Conclusions

This paper presents an ecological-inspired algorithm for optimization, called ECO. The proposed algorithm uses cooperative search strategies where populations of individuals co-evolve and interact among themselves using some ecological concepts. Each population behaves according to the mechanisms of intensification and diversification, and the control parameters, specific to a given search strategy. In this work we used the Artificial Bee Colony Optimization algorithm and the Particle Swarm Optimization algorithms to build an heterogeneous/hybrid ecological-inspired algorithm.

The main ecological concepts addressed are the definition of habitats, ecological relationships and ecological successions. Along with these concepts, the definition of intra and inter-habitats communication topologies are included to compose the algorithm. These features bring a higher biological plausibility to the proposed algorithm.

Experiments were carried out considering the implementation of the algorithm with the use of ecological concepts and without the use of ecological concepts (application of stand alone algorithms and application of co-evolution in a ring topology). Four benchmark functions were used with different degrees of complexity. The results showed that the use of habitats and ecological relations influence significantly the co-evolution process of populations, leading to better solutions (when compared to the results not using the ecological concepts). Also, the use of different algorithms inside the ECO framework improved considerably the results.

This work is still in development and as future work we intend to analyze the influence of the control parameters (number of ecological successions, evolutive period, number of populations, and ρ threshold for creation of habitats) on the quality of solutions, as well as to add other search strategies in the proposed model. Currently, in order to bring more biological plausibility to the system other ecological concepts are being modeled.

References

1. de Castro, L.N.: Fundamentals of natural computing: an overview. Physics of Life Reviews 4(1), 1–36 (2007)
2. Engelbrecht, A.P.: Computational Intelligence: An Introduction, 2nd edn. Wiley, Chichester (2007)
3. Parpinelli, R.S., Lopes, H.S.: New inspirations in swarm intelligence: a survey. International Journal of Bio-Inspired Computation 3(1), 1–16 (2011)
4. Begon, M., Townsend, C.R., Harper, J.L.: Ecology: from individuals to ecosystems, 4th edn. Blackwell Publishing, Oxford (2006)
5. May, R.M.C., McLean, A.R.: Theoretical Ecology: Principles and Applications. Oxford University Press, Oxford (2007)
6. El-Abd, M., Kamel, M.: A Taxonomy of Cooperative Search Algorithms. In: Blesa, M.J., Blum, C., Roli, A., Sampels, M. (eds.) HM 2005. LNCS, vol. 3636, pp. 32–41. Springer, Heidelberg (2005)

7. Masegosa, A.D., Pelta, D., del Amo, I.G., Verdegay, J.L.: On the Performance of Homogeneous and Heterogeneous Cooperative Search Strategies. In: Krasnogor, N., Melián-Batista, M.B., Pérez, J.A.M., Moreno-Vega, J.M., Pelta, D.A. (eds.) NICSO 2008. SCI, vol. 236, pp. 287–300. Springer, Heidelberg (2009)

8. Parpinelli, R.S., Benítez, C.M.V., Lopes, H.S.: Parallel approaches for the artificial bee colony algorithm. In: Panigrani, B.K., Shi, Y., Lim, M. (eds.) Handbook of Swarm Intelligence: Concepts, Principles and Applications. Series: Adaptation, Learning, and Optimization, pp. 329–346. Springer, Berlin (2011)

9. Benítez, C.M.V., Parpinelli, R.S., Lopes, H.S.: Parallelism, hybridism and coevolution in a multi-level ABC-GA approach for the protein structure prediction problem. In: Concurrency and Computation: Practice and Experience (2011)

10. Parpinelli, R.S., Lopes, H.S.: An eco-inspired evolutionary algorithm applied to numerical optimization. In: Proceedings of the Third World Congress on Nature and Biologically Inspired Computing, Salamanca, Spain, pp. 473–478 (2011)

11. Karaboga, D., Akay, B.: A comparative study of artificial bee colony algorithm. Applied Mathematics and Computation 214, 108–132 (2009)

12. Clerc, M.: Particle Swarm Optimization. ISTE Press (2006)

13. Blickle, T.: Tournament selection. In: Bäck, T., Fogel, D., Michalewicz, Z. (eds.) Evolutionary Computation, vol. 2, pp. 181–186. Institute of Physics, Bristol (2000)

14. Digalakis, J.G., Margaritis, K.G.: An experimental study of benchmarking functions for evolutionary algorithms. International Journal of Computer Mathematics 79(4), 403–416 (2002)

15. Floudas, C.A., Pardalos, P.M.: A Collection of Test Problems for Constrained Global Optimization Algorithms. LNCS, vol. 455. Springer (1990)

16. Mühlenbein, H., Schomisch, D., Born, J.: The parallel genetic algorithm as function optimizer. Parallel Computing 17(6-7), 619–632 (1991)

17. Griewank, A.: Generalized descent for global optimization. Journal of Optimization Theory and Applications 34(1), 11–39 (1981)

18. Cho, H., Olivera, F., Guikema, S.: A derivation of the number of minima of the Griewank function. Applied Mathematics and Computation 204(2), 694–701 (2008)

19. Rosenbrock, H.: An automatic method for finding the greatest or least value of a function. The Computer Journal 3, 175–184 (1960)

20. Clerc, M.: Standard PSO 2007, SPSO-07 (2007),
http://www.particleswarm.info/Programs.html

Providing Trade-Off Techniques Subsets to Improve Software Testing Effectiveness: Using Evolutionary Algorithm to Support Software Testing Techniques Selection by a Web Tool

Aurélio da Silva Grande, Arilo Claudio Dias Neto, and Rosiane de Freitas Rodrigues

Instituto de Computação (IComp)
Programa de Pós-Graduação em Informática (PPGI)
Universidade Federal do Amazonas, Manaus, Amazonas, Brasil
{aurelio.grande,arilo,rosiane}@icomp.ufam.edu.br

Abstract. The combination of testing techniques is considered an effective strategy to evaluate the quality of a software product. However, the selection of which techniques to combine in a software project has been an interesting challenge in the software engineering field because the high number of techniques available at the technical literature. This paper presents an approach developed to support the combined selection of model-based testing techniques, applying multiobjective combinatorial optimization strategies, by determining the minimum dominating set in a bipartite and bi-weighted graph. Thus, an evolutionary strategy based on a multiobjective genetic algorithm is proposed to generate trade-off techniques subsets between the maximum coverage of software project characteristics and the minimum eventual effort to construct models used for test cases generation. In an empirical evaluation, our evolutionaryalgo-rithmstrategygavebetterresultsthanthepreviousapproaches.

1   Introduction

This work addresses a very important problem in the software engineering field: the selection of techniques to test a software project. In order to propose solutions for this problem, we applied combinatorial optimization and graph theory, modeling the problem as a variation of the classical set covering problem considering multiple criteria to optimize. Thus, we defined the problem of determining the minimum dominating set in bipartite and bi-weighted graphs, to model the problem. This is a NP-hard problem, where approximate algorithmic strategies are suitable to solve them in real applications, to get good solutions in a very fast computational time.

In this way, we are proposing a multiobjective genetic algorithm to solve this problem by generating sets of test techniques with a trade-off between maximum coverage of software projects characteristics and minimum eventual effort to construct models used for test cases generation. This GA evolutionary strategy, named *Porantim-GA*, was incorporated in a web tool to support the software testing techniques selection process.

The remainder of this paper is organized as follows. In Section 2, we present definitions and concepts relevant in Software Engineering concern to the software

L.N. Barros et al. (Eds.): SBIA 2012, LNAI 7589, pp. 222–232, 2012.

testing techniquesselection problem, with related previous works. A graph theoretical model and combinatorial optimization concepts are given in Section 3. In Section 4, we present our proposed method, a genetic algorithm that generates a population of chromosomes evaluated for more than one fitness function, generating trade-off solutions. In Section 5, we present a comparative analysis of testing techniques selection approaches. Finally, in Section 6, we present concluding remarks.

2   Problem of Software Testing Techniques Selection (STTS)

In the context of software engineering, software testing consists in a technique that executes software revealing failures before its deployment to the end user and providing information regarding its quality in the environment where it will be executed [1]. The combination of testing techniques would be an interesting option to obtain success in the testing activities.

The combination of testing techniques means to apply different strategies to support testing for a same software product with the purpose of complement eventual limitations of a technique and thus improve the testing quality (testing effectiveness). However, to increase testing effectiveness may also require greater effort or cost for testing when the selected testing techniques have conflicting requirements to be used in the same software project. Therefore, the testing techniques selection needs to be carefully analyzed, as they may render the testing activities in a project unfeasible [2].

Some approaches have been proposed in the technical literature to support the selection of testing techniques with different characteristics [3,4,5]. However, these selection approaches do not provide supporting to the combined selection of testing techniques for a same project. This scenario motivated the development of a selection approach supportingthecombinedselectionoftestingtechniques,calledPorantim[6].

Porantim is a testing techniques selection approach based on two main elements:

1. **Body of Knowledge on MBT Techniques:** a repository of testing techniques identified from a literature systematic review described in [7]. Currently, it is composed of 219 different and real testing techniques.

2. **MBT Techniques Selection Process:** it provides information regarding the adequacy of the selected techniques for a software project. The adequacy of a testing technique is calculated in comparison with the software project's characteristics.Itcorrespondstoanumericalindicatorcalculatedwithformulas,descr ibed in [7], that explore the mathematical concept of Euclidean Distance.

Results of experimental studies revealed *Porantim* would be more efficient and effective than other selection approaches [6]. However, providing the combined selection of testing techniques using *Porantim* requires a high effort to combine manually the testing techniques suggested individually by the approach. Thus, the generation of best options (testing techniques combinations) regarding which testing techniques would be best suited for a software project is still dependent of which techniques were pre-selected by the testing team, because would be unfeasible for a human evaluate all possible combination of techniques. In order to deal to this limitation, *Porantim* has been evolved using a Multiobjective Combinatorial Optimization strategy to implement construction heuristics aiming at finding good solutions for the testing techniques selection problem in a very fast way.

The next section will present an evolution of *Porantim* implementing a multiobjective combinatorial optimization strategy to support the selection of testing techniques in software project.

3   A Graph Theoretical Model for STTS Problem

The problem of combined selection of software testing techniques can be modeled as a specific case of the classical set covering problem (SCP), using a model in graphs, know in the literature as the minimum dominating set problem (DSP), as described below. Both are classic problems in computer science and complexity theory, which have been intensely researched.

The DSP in general graphs is a NP-hard problem [8]. It remains NP-hard even in bipartite graphs. The graph can be weighted, that is, each edge of the graph can has a weight, so, to optimize this problem is quite different. However, the model of the DSP in graphs that best represents the STTS problem is even more complex than these previous versions, because we must consider two weights simultaneously on the edges of the graph. So, the DCP multiobjective version for model the STTS problem, can be now stated as:

Minimum Dominating Set in bipartite and bi-weighted graphs: given a graph $G=(V,E)$ with a bipartition $V_1, V_2 \in V$ and two weight w_1 and w_2 associated to each edge $e=\{v1,v2\}$, where $v_1 \in V_1$ e $v_2 \in V_2$, determine the smallest number of vertices – the dominating set $D \in V_1$ in G, such that for all remain vertex $v \in V_2$ there is a vertex $u \in D$ for which the edge $\{u,v\} \in E$, and the total sum of its weights is the best possible compromise.

Thus, given a set of $n_1 = |V_1|$ testing techniques (first partition of the graph) and a software project to be assessed by these techniques, where this project has $n_2 = |V_2|$ attributes in question (the second partition of the graph), the problem consists in to find a subset D of testing techniques that "dominate" or "cover" all the attributes of the project under evaluation, such that thissubset of testing techniques have the best compromise between maximum coverage and minimal effort of the attributes of construction or adaptation of models to be used for generation of tests (Fig. 1).

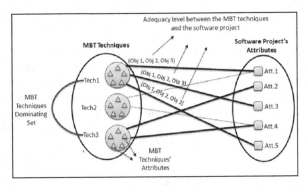

Fig. 1. Modeling of STTS problem as a bipartite and bi-weighted graph

Given the problem, we want to get good trade-offs (subset of testing techniques with the best compromise values) between the two objectives involved, that is, ideally we want to get the optimal Pareto of the problem, which constitutes the best compromise solutions or non-dominated solutions region of feasible solutions of the problem with two objectives to optimize [9,10].

This multiobjective version of the DSP is also NP-hard, as the previous versions, but with an even bigger factor of complexity. Therefore, generate the minimum dominating set accurately has a high computational cost. In practice, can be possible that this dominating set does not exist, that is, can be possible there is not exist a dominating set of testing techniques that covering exactly all the attributes of the project and that, moreover, minimize the effort of building models (in other words, can be impossible to have techniques that meet all the characteristics defined for the software project in question and still only adopt models already provided by the process of software development organization, without requiring an extra effort in their construction for generation of tests).

In this way, it is well suitable to adopt approximate algorithmic approaches, seeking to generate a subset of testing techniques that not necessarily cover all the attributes of the project, but satisfying a relation between maximum coverage with minimal effort, since the objective of *Porantim* approach is just provide information about different possibilities for combining testing techniques which is relevant to a software project to aid in a decision making process to be performed by the test team, and not make the decision instead. A strong condition defined for any approximate algorithmic approach is that the results are computed with a reasonable computational time, because this decision must be made during the course of a software development project, and there would be no time or resources available for implementing algorithms with high computational cost.

The next section shows the method proposed in this paper, a genetic algorithm to solve the multiobjective version of the problem as a natural extension of the effort to solve this problem keeps getting better, and thus perform a suitable theoretical and practical application between Combinatorial Optimization and Software Engineering.

4   *Porantim-GA:* Selecting Testing Techniques with Multiobjective Genetic Algorithm

Initially, the evolution of *Porantim* was done by implementing a local search strategy. This approach was called *Porantim-Opt*. In this approach, a multiobjective strategy was used because the adequacy level indicator, used by *Porantim*, was spitted in two objectives: characteristics coverage and modeling effort. Using *Porantim-Opt*, new solutions are generated considering three steps: (1) Calculating individual analysis of testing techniques; (2) Combining two current solutions in a greedy heuristic strategy, and (3) Repeating the process until the stop criteria is reached.

After that, we decided to implement a solution to the STTS problem using an evolutionary strategy embedded in a multiobjective genetic algorithm (GA). A GA is a multi-solution or population-based metaheuristic that is meant to solve the problem by the evolution of a set of initial solutions, by generating successive sets of solutions in an iterative process that exploits the space of feasible solutions enabling a balance between ensuring diversity and convergence.

In our case, we developed a multiobjective GA that uses the front of Pareto like strategy in which the first front of Pareto of the current generation is taken to the next generation. At the end, a dominance-based approach is applied for the final evaluation of solutions [9,10].The pseudocode of the algorithm implemented in *Porantim-GA*is described in Fig. 2.

```
1. Initialize the initial population of individuals.
2. Repeat on this generation until stopping criteria is reached:
   2.1. Select the best-fit individuals (trade-off solutions considering the two
        objectives) for reproduction
   2.2.Breednewindividualsthroughcrossoverandmutationoperations. (each one with
        50% of population without the individuals selected on step 2.1)
   2.3. Replace least-fit population with new individuals.
3. Select the trade-off solutions obtained in the last generation.
```

Fig. 2. Porantim-GA's pseudocode

In the next subsections we present the details of this selection approach.

4.1 Chromosome and Candidate Solutions

A chromosome corresponds to a solution of the problem.Itis formed by combining one, two, or three elements, that are testing techniques in our case. Therefore, the size of the chromosome is dynamic containing one, two or three genes (elements), each one with an integer number as value (allele) that represents a specific testing technique, as presented in Fig. 3. The position of an element in a chromosome (locus) is not important in this case.Moreover, the same technique can appear in more than one chromosome. A chromosome is invalid when atechniqueappearsrepeatedly.

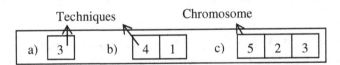

Fig. 3. Representation of the chromosomes with (a) 1, (b) 2, (c) or 3 elements

4.2 Objectives

The two objectives to be optimized (maximized) in the problem are described below. They were originated in the original version of *Porantim* [6]:

- *SWP_Cov* (**Software Project Coverage**) = indicator of how many software project's characteristics are being covered by the selected testing techniques.
- *SavModEffort* (**Saved Modeling Effort**) = indicator of how many models required to use the selected techniques are already provided (and how) in the software development process, saving effort to construct test models.

4.3   Initialization, Selection, Reproduction, and Stopping Strategies

In order to generate an initial population ensuring high diversity, all testing techniques are grouped randomly in solutions containing three techniques. Thus, all of them are candidates to be part of solutions that will be derived from the initial population.

For selection, each generation of the genetic algorithm has 50 different individuals or solutions (population size). The first front of Pareto is used to select the non-dominated solutions – elitism.

The reproduction strategy is composed of two operations: <u>crossover</u> and <u>mutation</u>. (Each one with 50% of population size without the selected elements of elitism) The first one is based on the combination of two solutions chosen randomly from the previous population. The combination depends on the size of the selected solutions, whichcangenerateanewsolutionwithtwoorthreetechniques,asdescribedonFig.4.

```
Crossover Strategy
  → Give the actual Population:
  1.    T1 = Select one individual (solution) randomly;
  2.    T2 = Select another individual randomly;
  3.    L1 = length of T1;
  4.    L2 = length of T2;
  5.    New = Left side of T1 (1 or 2 techniques) ∪ Right side of T2
(1 or 2 techniques);
  6.    Check if New is valid and it does not exist in the new popu-
lation;
```

Fig. 4. Algorithm that implements the Crossover Strategy (subpart of Reproduction Strategy)

The second operation (mutation) modifies randomly an existing solution. If the chromosome is composed of only one technique, one or two techniques are selected randomly and added to form a new solution. Otherwise, one technique is excluded from the selected solution.

Finally, the stopping criteria are also two: after the production of 100 generations OR when the five best solutions are the same (there is no more evolution, reaching convergence) past ten consecutive iterations.

4.4   Trade-Off Solutions Selection and the Computational Infrastructure

That is a possibility offered by the multiobjective analysis used in the proposed approach. Inthe end of each iteration the dominance analysis (Pareto front with the non-dominated solutions that maximize the objectives in question) is performed with the two objectives in order to select individuals to form the new generation. In the last iteration, the Pareto front will represent the final solutions provided by the algorithm.

In order to support the use of *Porantim-GA*approach and its old versions (*Porantim* and *Porantim-Opt*), their algorithms were implemented in the Maraká tool[1]. Maraká is

---

[1] www.icomp.ufam.edu.br/maraka

a web-based computational infrastructure to support test management, from planning to execution of the tests, following a process of pre-defined test[11]. It is available for software companies, because it has a number of features,suchasgraphical monitoring of tests, charts of revealed failures, generation tests documentation, management of testers tasks, among other features. It also provides a feature to support the selection of testing techniques for a software project. *Porantim-GA*'s (and its old versions) features were implemented in this part.

Porantim-GA's operation in the Maraká tool consists of three steps: (1) characterization of software projects through its attributes; (2) selection of techniques appropriate to the project (Fig. 5); and (3) confirmation of the performed selection.

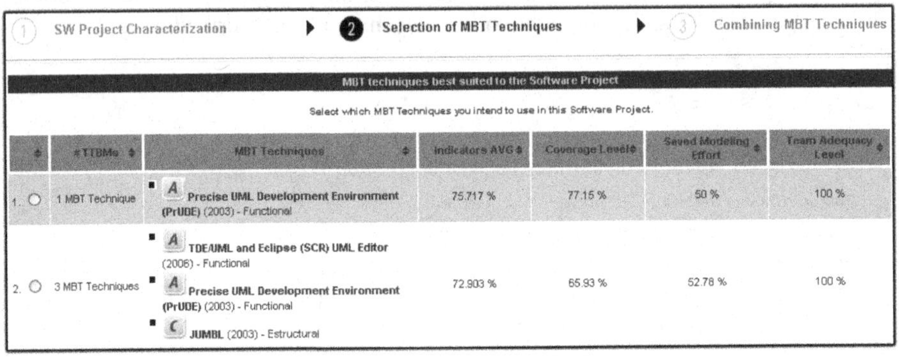

	TTBMs	MBT Techniques	Indicators AVG	Coverage Level	Saved Modeling Effort	Team Adequacy Level
1. ○	1 MBT Technique	▪ A Precise UML Development Environment (PrUDE) (2003) - Functional	75.717 %	77.15 %	50 %	100 %
2. ○	3 MBT Techniques	▪ A TDE/UML and Eclipse (SCR) UML Editor (2006) - Functional ▪ A Precise UML Development Environment (PrUDE) (2003) - Functional ▪ C JUMBL (2003) - Estructural	72.903 %	65.93 %	52.78 %	100 %

Fig. 5. Selection of MBT Techniques' Maraká Screen

In the next section, it will be described a comparative analysis among *Porantim-GA* and its previous versions (*Porantim* and *Porantim-Opt*).

5   Comparative Analysis of the Selection Approaches

The conducted experiment had the goal of comparing the effectiveness of three selection approaches (*Porantim, Porantim-Opt*, and *Porantim-GA*) for two projects previously selected: <u>Video Manager</u> and<u>Parking System</u>. These projects were defined in a package published in [4] and they were used in the evaluation of the original version of *Porantim*.

The three selected approaches have some differences. The first one, *Porantim*, implements an algorithm that does not use optimization resources and it does not provide solutions with combination of techniques. The second one, *Porantim-Opt*, implements a greed metaheuristic that combines techniques to reach new enhanced solutions. However, as a greed algorithm, the space of solutions is not exploited totally and there are others metaheuristics in the literature that have better performance. Furthermore, both approaches were not implemented as a multiobjective algorithm. Its fitness function considers the average of all objectives,individually calculated in a previous step. The third approach, *Porantim-GA*, implements a multiobjective evolutionary algorithm dealing separately to each objective. Then, it calculates the non-dominated solutions in each generation.

As benchmark for this experiment, we used the repository of testing technique provided in Maraká that represents the body of knowledge of *Porantim*. It contains 219 different and real testing techniques identified in the technical literature. If analyzed manually combining 3 testing techniques, this repository makes possible the instantiation of more than 10.000.000 solutions (219 x 218 x 217). Thus, the selection approaches were evaluated simulating their application in real and complex scenarios.

Each software project was analyzed individually. Thus, each selection approach was executed collecting the two goals (software project coverage [SPC] and saved modeling effort [SME]). Using *Porantim* and *Porantim-Opt*, only one execution was performed, because the results generated are always the same for a software project. In the case of *Porantim-GA*, we decided to run the algorithm 10 times for each project and then we selected the best and most recurrent results. 50 solutions were generated using each approach (some of them have repeated results).

5.1 Analyzing the Software Project Video Manager

Fig. 6 presents the results obtained considering all three selection approaches evaluated in this research for the software project **Video Manager**. At total, seven points (solutions) compose the Pareto Front.

Analyzing the *Porantim* results, its solutions (represented by a star in Fig. 6) are mainly concentrated in the middle of the chart, being dominated by other solutions. However, it generated only one solution non-dominated by the other selection approaches (*Porantim-Opt* and *Porantim-GA*), tagged as "F". This solution is composed of only one technique and it obtained SPC = 77.78% and SME = 58.33%.

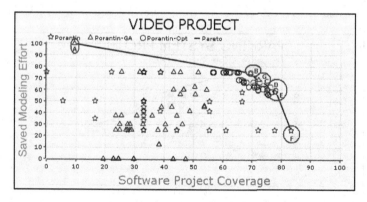

Fig. 6. Pareto Front with trade-off solutions combining the 3 approaches for Video Project

Analyzing *Porantim-Opt*, we can observe its solutions (represented by circles) are close to the best ones and concentrated near the Pareto front. Moreover, it generated four non-dominated solutions, being one composed of only one technique and three solutions composed of three techniques each one. They are tagged in Fig. 6 as "B" (SPC = 69.56% and SME = 75%), "C" (SPC = 72.65% and SME = 66.67%), "D" (SPC =76.32% and SME=62.50%) and "E" (SPC=78.04% and SME=59.37%).

Finally, *Porantim-GA* approach generated solutions (represented by triangle) more scattered on the graph, exploring the solution space. It still generated three non-dominated solutions. One of themis the same solution generated by *Porantim-Opt* (tag "B"), and two are new non-dominated solutions, one composed of 1 techniques and other composed of 2 techniques. They are tagged as "A" (SPC = 9.4% and SME = 100%) and "G" (SPC = 68.75% and SME = 72.31%).

5.2   Analyzing the Software Project Parking System

Analyzing the second project, Parking System, Fig. 7 presents the results obtained considering all three selection approaches evaluated in this research. At total, five points (solutions) compose the Pareto Front.

Porantim approach did not obtained non-dominated solutions for this software project.Itssolutions(representedbystars)aremainlyconcentratedinthemiddleofthe chart, being dominated by solutions generated by the other two selection approaches.

Analyzing *Porantim-Opt*, we can observe its solutions (represented by circles) are close to the best ones and concentrated near the Pareto front. Moreover, it generated 3 non-dominated solutions, being 2 composed of only 1 technique, tagged as "A" (SPC = 23.14% and SME = 100%) and "E" (SPC = 77.15% and SME = 50%) and 1 solution composed of 2 techniques tagged as "B" (SPC = 58.83% and SME = 75%).

Finally,*Porantim-GA* approach generated solutions (represented by triangle) again more scattered on the graph, exploring the solution space. This approach generated three non-dominated solutions. One of them is the same solution generated by *Porantim-Opt* (tag "A"), and two are new non-dominated solutions, one composed of 3 techniques tagged as "C" (SPC = 61.6% and SME = 62.5%) and other composed of 2 techniques tagged as "D" (SPC = 77.15% and SME = 50%). They are tagged as "A" (SPC = 9.4% and SME = 100%) and "G" (SPC = 68.75% and SME = 72.31%).

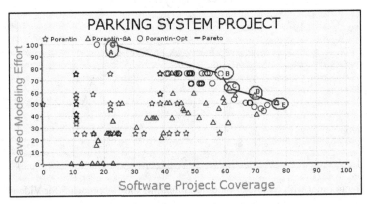

Fig. 7. Pareto Front with trade-off solutions combining the 3 approaches for Parking System

In the next section it will be described the conclusions of this paper, with ongoing and further works.

6   Conclusions

This paper introduced an approach to support the selection of testing techniques in software projects based on an evolutionary strategy by genetic algorithm. It was presented a theoretical model for Software Testing Techniques Selection Problem as a graph optimization problem. The evolutionary algorithm was implemented in a supporting web tool Maraká [11] that contain a set of 219 real testing techniques collected in the technical literature and it has been used as a benchmark.

Computational testswere performed in order to compare the new proposed selection approach (*Porantim-GA*), to its previous versions, that do not implement any evolutionary heuristic approach (*Porantim* and *Porantim-Opt*), neither trade-off solutions. In general, the first approach, *Porantim*, had the worst results, even though it achieved non-dominated solutions in one project, but on the second project it was overcome by the other two approaches. With respect to approaches *Porantim-Opt* and *Porantim-GA*, both generated non-dominated solutions for all projects, but *Porantim-GA* generated most trade-off solutions for the second project.

Considering all solutions, *Porantim-Opt* generated solutions very close to the best Pareto front generated by the multiobjective GA (*Porantim-GA*). Another issue to be considered is that *Porantim-GA* implements a non-deterministic algorithm. Thus, it presents different solutions each time it is executed. Finally, as verified in the experiment, *Porantim-GA* gave good results when compared to other approaches.

As ongoing works, we intend to execute new experiments with *Porantim-GA* considering new scenarios (software projects) and comparing with different selection approaches. Moreover, we are extended the idea of *Porantim-GA* to support the selectionofsoftwaretechnologies in general (techniques,processes,methods,andtools) for indifferentactivities,such as requirement elicitation, software design, and project management. Iterated local search method and hybrid evolutionary strategies will be considered in future works.

Acknowledgments. The authors acknowledge the support granted by FAPEAM, CNPq and FAPESP to the INCT-SEC (National Institute of Science and Technology – Critical Embedded Systems – Brazil), processes 573963/2008-8 and 08/57870-9.

References

1. Juristo, N., Moreno, A., Vegas, S.: Reviewing 25 years of testing technique experiments. Empirical Software Engineering: An International Journal 9(1), 7–44 (2004)
2. Menzies, T., Owen, D., Cukic, B.: —Saturation Effects in Testing of Formal Models ‖ . In: 13th international Symposium on Software Reliability Engineering (ISSRE 2002), Washington, DC, p. 15 (2002)
3. Wojcicki, M.A., Strooper, P.: An Iterative Empirical Strategy for the Systematic Selection of a Combination of Verification and Validation Technologies. In: Proceedings of the 5th International Workshop on Software Quality, May 20-26 (2007), http://dx.doi.org/10.1109/WOSQ.2007.4
4. Vegas, S., Basili, V.: A Characterization Schema for Software Testing Techniques. Empirical Software Engineering 10(4), 437–466 (2005)

5. Victor, M., Upadhyay, N.: Selection of Software Testing Technique: A Multi Criteria Decision Making Approach. In: Nagamalai, D., Renault, E., Dhanuskodi, M. (eds.) CCSEIT 2011, Part 1. CCIS, vol. 204, pp. 453–462. Springer, Heidelberg (2011)
6. Dias-Neto, A.C., Travassos, G.H.: Model-based Testing Approaches Selection for Software Projects. Information and Software Technology 51, 1487–1504 (2009)
7. Dias-Neto, A.C., Travassos, G.H.: A Picture from the Model-Based Testing Area: Concepts, Techniques, and Challenges. Advances in Computers 80, 45–120 (2010)
8. Papadimitriou, C., Steiglitz, K.: Combinatorial Optimization: Algorithms and Complexity. Chapman & Hall and CRC Press, New York (2004)
9. Talbi, E.: Metaheuristics: from design to implementation. Wiley Series on Parallel and Distributed Computing series. John Wiley & Sons (2009)
10. Abraham, A., Jain, L., Goldberg, R.: Evolutionary Multiobjective Optimization. Theoretical Advances and Applications. Springer, USA (2005)
11. Dias-Neto, A.C., Travassos, G.H.: Evolving a Computerized Infrastructure to support the Selection of Model-Based Testing Techniques. In: IFIP International Conference on Testing Software and Systems (ICTSS 2010), Brazil (2010)

Author Index